Gender, Agriculture and Agrarian Transformations

This book presents research from across the globe on how gender relationships in agriculture are changing.

In many regions of the world, agricultural transformations are occurring through increased commodification, new value chains, technological innovations introduced by CGIAR and other development interventions, declining viability of small-holder agriculture livelihoods, male out-migration from rural areas, and climate change. This book addresses how these changes involve fluctuations in gendered labor and decision making on farms and in agriculture and, in many places, have resulted in the feminization of agriculture at a time of unprecedented climate change. Chapters uncover both how women successfully innovate and how they remain disadvantaged when compared to men in terms of access to land, labor, capital and markets that would enable them to succeed in agriculture. Building on case studies from Africa, Latin America and Asia, the book interrogates how new agricultural innovations from agricultural research, new technologies and value chains reshape gender relations.

Using new methodological approaches and intersectional analyses, this book will be of great interest to students and scholars of agriculture, gender, sustainable development and environmental studies more generally.

Carolyn E. Sachs is Professor Emerita of Rural Sociology and Women's, Gender, and Sexuality Studies, Penn State University, USA.

Earthscan Food and Agriculture Series

The Financialization of Agri-Food Systems
Contested Transformations
Edited by Hilde Bjørkhaug, André Magnan and Geoffrey Lawrence

Redesigning the Global Seed Commons
Law and Policy for Agrobiodiversity and Food Security
Christine Frison

Organic Food and Farming in China
Top-down and Bottom-up Ecological Initiatives
Steffanie Scott, Zhenzhong Si, Theresa Schumilas and Aijuan Chen

Farming, Food and Nature
A Sustainable Future for Animals, People and the Environment
Edited by Joyce D'Silva and Carol McKenna

Governing Sustainable Seafood
Peter Oosterveer and Simon Bush

Farming Systems and Food Security in Africa
Priorities for Science and Policy Under Global Change
Edited by John Dixon, Dennis P. Garrity, Jean-Marc Boffa, Timothy Olalekan Williams, Tilahun Amede with Christopher Auricht, Rosemary Lott and George Mburathi

Consumers, Meat and Animal Products
Policies, Regulations and Marketing
Terence J. Centner

Gender, Agriculture and Agrarian Transformations
Changing Relations in Africa, Latin America and Asia
Edited by Carolyn E. Sachs

For further details please visit the series page on the Routledge website: www.routledge.com/books/series/ECEFA/

Gender, Agriculture and Agrarian Transformations

Changing Relations in Africa, Latin America and Asia

Edited by Carolyn E. Sachs

Routledge
Taylor & Francis Group
LONDON AND NEW YORK

earthscan
from Routledge

First published 2019
by Routledge
2 Park Square, Milton Park, Abingdon, Oxon OX14 4RN

and by Routledge
605 Third Avenue, New York, NY 10017

First issued in paperback 2020

Routledge is an imprint of the Taylor & Francis Group, an informa business

British Library Cataloguing-in-Publication Data
A catalogue record for this book is available from the British Library

Library of Congress Cataloging-in-Publication Data
Names: Sachs, Carolyn E., 1950- editor.
Title: Gender, agriculture and agrarian transformations : changing relations in Africa, Latin America and Asia / edited by Carolyn E. Sachs.
Description: New York, NY : Routledge, 2019. | Series: Earthscan food and agriculture series | Includes index.
Identifiers: LCCN 2019002446 (print) | LCCN 2019010865 (ebook) | ISBN 9780429427381 (Master) | ISBN 9781138384941 (hardback) | ISBN 9780429427381 (ebook)
Subjects: LCSH: Women in agriculture—Africa—Case studies. | Women in agriculture—Latin America—Case studies. | Women in agriculture—Asia—Case studies.
Classification: LCC HD6077 (ebook) | LCC HD6077 .G44 2019 (print) | DDC 338.1082—dc23
LC record available at https://lccn.loc.gov/2019002446

ISBN 13: 978-0-367-72855-7 (pbk)
ISBN 13: 978-1-138-38494-1 (hbk)

Typeset in Goudy
by Apex CoVantage, LLC

Contents

Figures

Tables

Contributors

Lemlem Aregu is a gender specialist located in Ethiopia who has worked extensively in Africa and Asia. She led gender research for WorldFish Myanmar from 2015 to 2018.

Diksha Arora is a Gender Economist at the International Center for Tropical Agriculture (CIAT). Her research interests gravitate around the issues related to intra-household dynamics, climate change and food security in the smallholder agricultural system with a regional focus on Latin America and sub-Saharan Africa.

Jacqueline A. Ashby is a Development Sociologist (Ph.D. Cornell University) who has worked in CGIAR senior management on the application of social and gender analysis to plant breeding and natural resource management, most recently as Senior Adviser for Gender in Research to the CGIAR System Office and the Gender and Breeding Initiative.

Luis Barba-Escoto is pursuing an M.Sc. in complexity sciences. He currently works as a research assistant with the socioeconomics and sustainable intensification programs at Centro Internacional de Mejoramiento de Maíz y Trigo (CIMMYT).

Edward Bikketi is a Gender Postdoc Fellow in the CGIAR Gender Research Program of ICRISAT since 2016 on Gender Gaps in Dryland Cereals and Grain Legumes. With a transdisciplinary background, he is promoting gender integration in breeding initiatives from ICRISAT's sustainable Grain Legumes Dryland Cereals CRP strategy.

Juan Burgueño-Ferreira is a senior scientist, Head of the Biometrics and Statistics Unit at the Centro Internacional de Mejoramiento de Maiz y Trigo (CIMMYT). He has been professor in the Universidad de la República, Uruguay, and the Colegio de Postgraduados, México, and since 1999 has been working at CIMMYT. His research interests are the biometrical and statistical methods applied in natural sciences.

Tania Carolina Camacho-Villa is an associate scientist in the Socioeconomics Program of CIMMYT. With a multidisciplinary background, she has been studying topics such as traditional farming systems, agrobiodiversity, innovation processes and scaling. Since 2015, she is promoting social inclusion processes in projects from CIMMYT Sustainable Intensification Strategy in Latin America, taking into account gender, age and cultural inclusion.

Afrina Choudhury is a gender specialist at WorldFish Center in Dhaka, Bangladesh with research interests on finding ways to integrate gender into technical interventions in ways that are sustainable and transformative.

Mónica Clavijo is an economist with master's studies in environment and development and is based at the Environmental Studies Institute at the National University of Colombia.

Todd A. Crane is an ecological anthropologist and a senior scientist at ILRI working on climate change adaptation and mitigation issues.

María del Rosario Castro Bernardini is a rural sociologist and women's studies scholar who teaches women's, gender and sexuality studies at Penn State.

Sujata Ganguly is an independent consultant with special interest on issues related to gender, migration and health.

Leland Glenna, Professor of Rural Sociology and Science, Technology, and Society, examines the social, economic, environmental, ethical and policy effects of agricultural, food and environmental science and technology.

Tatiana Gumucio is an anthropologist and Gender Specialist in the CGIAR Research Program on Climate Change, Agriculture and Food Security (CCAFS), based at the International Research Institute for Climate and Society.

Leif Jensen is Distinguished Professor of Rural Sociology and Demography at Penn State, where he studies social inequality, underemployment, informal work, gender and agriculture and rural people and places domestically and internationally.

Edna Johnny is a Ph.D. student at the University of Nairobi School of Economics.

Nozomi Kawarazuka is Associate Scientist at the social and nutritional sciences division, International Potato Center (CIP) based in Vietnam. Her research focuses on gender and agricultural innovation and scaling, ethnic minorities, labor migration, food systems and culinary innovation.

Stephanie Leder is Postdoctoral Researcher at the Swedish University of Agricultural Sciences (SLU) interested in feminist political ecology, water resource management, critical gender and development research and Education for Sustainable Development in India, Nepal and Bangladesh.

Santiago López-Ridaura is a senior scientist at the Centro Internacional de Mejoramiento de Maiz y Trigo (CIMMYT), based in its headquarters in Mexico.

His research interests are the integrated assessment of farming systems and the use of modeling tools to explore scenarios and understand trade-offs among different objectives. He has more than 15 years of experience in the development of approaches and tools for farming systems analysis.

Cynthia McDougall is the Gender Research Leader for WorldFish and the CGIAR Research Program on Fish Agri-food Systems. She is an interdisciplinary social scientist with more than 20 years of experience in food security, gender and social equity and natural resource governance.

Birhanu Megersa Lenjiso is an agricultural Sociologist and a Postdoctoral Scientist, working on gender and agricultural technology adoption at the International Livestock Research Institute (ILRI).

Samarendu Mohanty is Asia regional director for the International Potato Center (CIP) in Vietnam. He is an agricultural economist with skills in commodity markets and policy and trade analysis.

Sugandha Munshi is a gender specialist at the International Rice Research Institute in Bihar, India. She closely works with women farmers on innovation and policy and has worked to establish the identity of women farmers in the state of Bihar.

Swati Nayak is a social scientist at the International Rice Research Institute in New Delhi working in designing, implementing various socioeconomic research, partnership development around social and gender inclusion, training and capacity-building activities. She has experience with various organizations engaged in gender and social inclusion, primary livelihood promotion, conservation and sustainable agriculture promotion.

Esther Njuguna-Mungai is a Senior Gender Scientist at International Crops Research Institute for the Semi-Arid Tropics (ICRISAT) in Nairobi, Kenya.

Vivian Polar is an agronomist and a social scientist who has worked extensively with women and indigenous peoples in developing countries on empowerment for agricultural innovation, most recently as Gender and M&E specialist for the CGIAR Crop Research Program on Roots, Tubers and Bananas and the Gender and Breeding Initiative.

Surendran Rajaratnam is a senior research scientist at WorldFish, where he conducts a range of gender studies using gender-transformative approaches and tools.

Carolyn E. Sachs is Professor Emerita of Rural Sociology and Women's, Gender, and Sexuality Studies at Penn State University. She conducts research on gender and agriculture, sustainable agriculture and gender and environment.

Arindam Samaddar is an agricultural anthropologist working with the International Rice Research Institute since 2013 and is currently Lead Specialist-Monitoring Evaluation based in New Delhi.

Kathleen Sexsmith, Assistant Professor of Rural Sociology and Women's, Gender, and Sexuality Studies at Penn State University, works on community integration in Latinx communities in the rural U.S. and on the gender dynamics of voluntary sustainability standards for agriculture in the Global South.

Katie Tavenner, Ph.D., is a postdoctoral fellow in gender research at the International Livestock Research Institute (ILRI) in Nairobi, Kenya.

Nguyen Thi Van Anh is a demographer, senior researcher and gender specialist at the Institute for Social Development Studies (ISDS). She has engaged in various studies and training related to gender and women, reproductive health and stigma issues working with HIV/AIDS key affected populations and other vulnerable groups such as poor men and women, people with disabilities, LGBT, drug users and sex workers.

Pham Huu Thuong is an Agricultural Researcher at the World Agroforestry Centre (ICRAF) in Hanoi, Vietnam. With environmental and agricultural background, he has been working in agriculture, agroforestry topics with ethnic minorities since 2011 in mountainous areas in Vietnam.

Ann R. Tickamyer, Professor of Rural Sociology at Penn State University, has research interests in poverty, livelihood practices and welfare provision in rural Appalachia and Indonesia.

Jennifer Twyman is an agricultural economist and Gender Research Leader at the International Center for Tropical Agriculture (CIAT) based in Cali, Colombia.

Margreet van der Burg is a university lecturer and researcher in gender studies/history with focus on food, agricultural and rural research and development, Wageningen University, the Netherlands.

Prakashan Chellattan Veettil is an agricultural economist at IRRI in New Delhi, India, and conducts research on seed systems, rural entrepreneurship development for service economy, youth in agriculture and varietal adoption and technology impact assessment.

Vu Xuan Thai is Social Science Researcher at the Institute for Social Development Studies (ISDS) in Hanoi, Vietnam. He has 12 years' working experience in gender and development projects, working closely on issues of gender equality and women's rights.

Acknowledgments

This book is made possible through the joint efforts of CGIAR gender specialists and postdoctoral researchers and the Gender Equity Through Agriculture Research and Education (GEARE) team at Penn State University. The postdoctoral training program was the brainchild of Jacqui Ashby, who we would like to thank for her inspiration and participation. We engaged in a vibrant exchange about feminist theory, methods, gender approaches to development, working in collaboration with agricultural scientists and much more over a three-year period. We would like to thank Peter Gardiner at CGIAR for his continuing support of gender researchers to see us through the publication of this book. I would also like to thank members of the GEARE team at Penn State who have worked hard to inspire important and innovative scholarship on gender and agriculture including Mary Barbercheck, Ricky Bates, Deanna Behring, Kathy Brasier, Ty Butler, Paige Castellanos, Leland Glenna, Michael Jacobsen, Leif Jensen, Ruth Mendum, Patty Neiner, Edwin Rajotte, Elizabeth Ransom, Kathleen Sexsmith and Ann Tickamyer. A number of graduate and undergraduate students also helped support the work of GRIT including Meredith Field, María del Rosario Castro Bernardini, Elisabeth Garner, Sarah Fusco, Amani Abdullah Alkhiami and Blair Cooper. Also, Margareet van den Burg generously gave her time to participating in two years of the GRIT training. I would especially like to thank Deanna Behring for her enthusiastic and continuous support of the GEARE effort. Thanks to Robin Becker for her comments on the introductory chapter. Finally and of most importance, we would also like to acknowledge the hard work of the women and men working in agriculture, fisheries, forests, and environment in Africa, Asia, and Latin America who grow food for their families, communities, and beyond.

Part 1

Overview

1 Gender, agriculture and agrarian transformations

Carolyn E. Sachs

This book explores research from Asia, Africa and Latin America on shifting gender relationships in agriculture. Many factors affect agricultural transformation throughout the world. Our authors take up increased commodification, new value chains, and technological innovations introduced by agricultural development interventions. Other authors examine declining viability of smallholder agriculture livelihoods, male out-migration from rural areas and climate change. Because of these shifts in agriculture, women and men's participation in agri-food systems involves shifting gender roles and identities, which varies across different spaces and different types of agriculture. Fluctuations in gendered labor and decision making on farms and in agriculture recur throughout the book in sometimes contradictory and interesting ways. In some regions, these changes result in the feminization of agriculture, while in other places we witness the masculinization of agriculture. Almost everywhere, women remain disadvantaged compared to men in terms of access to land, labor, capital, and markets that enable them to succeed. Food insecurity at the household, community, regional and national levels results. While agriculture development efforts attempt to increase women's access to these resources to increase gender equity, they also aim to explicitly empower women. Several of the chapters in this volume uncover how women and men successfully innovate to develop more economic and environmentally sustainable agricultural systems to feed their households and communities. These emphasize women's creative agency and their strategic innovations. Several chapters interrogate agricultural innovations, new technologies and changes in value chains and how these reshape gender relations. New theoretical and methodological approaches go beyond the nuclear household as the unit of analysis and include intersectional analysis.

We seek to reveal how specific agricultural research and agricultural development efforts and policies facilitate or hamper gender equity and women's empowerment in a wide variety of agricultural systems in Africa, Asia and Latin America. As women take increasing responsibility for agriculture in many regions of the world, their ability to innovate and succeed depends on their positionality in the family and society. Gendered decision-making authority is deeply embedded in social and cultural history. Further, agricultural research and development strategies, such as value chain development or the development of new plant varieties, can facilitate or undermine gender equity.

These gender and agricultural scholars chart a course for the future of the field of gender and agriculture. The idea for this book emerged from an intellectually engaging summer workshop on Gender Research Intensive Training (GRIT) with gender researchers from the CGIAR and Penn State gender and agriculture faculty members. In the workshop, we discussed and debated emerging feminist epistemologies, strategies for conducting gender and agricultural research, collaborating across interdisciplinary teams and settings and new and innovative methodological approaches. The gender researchers from CGIAR presented their initial research findings, and we discovered we had a rich collection of field studies on gender and agriculture from Africa, Asia, and Latin America. We collaborated and worked on the papers over the next year and then met again for another intensive and exhilarating three-week session the next summer with returning and new gender researchers from CGIAR. At this stage, many of the papers incorporated ideas and methods from our workshop, and we had another chance to share, discuss and debate our work with each other. Our collaborations resulted in writing that addresses four critical areas in the field of gender and agriculture, including: (1) gender and agricultural research, (2) intersectionality: gender, class, and ethnicity in agricultural systems, (3) shifting gender relations in livestock and fisheries, and (4) gender, labor and decision making.

Much of the scholarship on gender and agricultural research has focused on the gaps in agricultural productivity and technology adoption between men and women (Udry, 1996; Burke and Banda, 2018). Most studies explain the gender yield gap as a result of uneven access to land, labor, inputs, technology, education, extension and markets. As issues of food security rise to the forefront of the international development agenda, opportunities to increase women's productivity offer hope for increasing the productivity of small-scale farmers. Many development efforts targeted toward increasing the productivity of women farmers focus on providing women with access to new technologies such as improved seeds and small-scale irrigation systems. However, important questions remain as to whether women and men actually desire different types of technology and, if so, why and how. Ashby and Polar explore whether men and women are likely to adopt the same technology, or if a diversity of technologies might better serve the interests of both men and women. Their study focuses on 39 case studies that assess if there are gender-differentiated preferences for particular traits in plants, such as length of growing season and cooking quality. They suggest that plant breeders do not necessarily need to develop varieties specifically for men and women but that they should strive to develop some varieties that contain the traits that women prefer. They pinpoint the moment in plant breeding development that breeders can consider gender trait preferences and provide a detailed guide for the gender screening of varietal traits. Van der Berg takes a historical look at gender research in the CGIAR. As the major international agriculture research system with 15 separate centers, CGIAR employs more than 8,000 staff located across the globe. She examines how strategies for incorporating gender into agricultural research at various CGIAR institutions changed over time, beginning in the 1970s. Her in-depth study carefully documents how paradigmatic shifts in CGIAR research such as farming systems and participatory action research created openings and opportunities for innovative gender research. However, she finds that gender research in the CGIAR ebbed and flowed and did not develop in a linear pattern toward more transformative gender research. In fact, many writers, emerging scholars within CGIAR system, seek to move towards more transformative gender research. Sexsmith et al. note the lack of attention to gender in agricultural research, beginning with the simple lack of sex-disaggregated data to the more complicated issues of understanding gender dynamics. These writers provide feminist epistemological and methodological strategies that can guide agricultural research, policy and practices.

Rather than simply focusing on the differences between men and women, intersectional analysis helps us understand the complexities of how gender intersects with class, ethnicity, age, and marital status in work and decision making. Central to gender studies, intersectional theories and analysis are less common and just gaining traction in gender and agricultural research. Leder and Sachs's study focuses on how efforts to study and provide outreach with the goal of women's empowerment must more carefully understand the complex intersections of multiple factors in women's lives. Based on fieldwork in Nepal, they show how complicated intra- and inter-household factors impact women's agricultural practices and decision making. Camacho-Villa et al. present an analysis of different types of households in the Western Highlands of Guatemala. This maize-growing region contains rich agrobiodiversity and social diversity but is also characterized by poverty and food insecurity. The study develops a typology of farm households that includes a gender dimension. This typology points out how increasing land fragmentation as well as the feminization of agriculture result in quite different needs for multiple types of maize-growing households. They suggest how agricultural development efforts might better target these types of households to reduce food insecurity. Finally, Kawarazuka et al. present a fascinating study of Hmong young women in the highlands of Vietnam, revealing how both young women and men struggle for autonomy in patriarchal intergenerational households. Here, young women do heavy manual labor in agricultural systems that focus on coffee, maize and local fruits. The authors show how some young women break with tradition to pursue innovative strategies in agriculture but others, in the words of one woman, feel like "a bird locked in a cage."

Gender has always been a factor in the raising of livestock, aquaculture and fisheries. Ownership of and access to livestock and fisheries in many regions contains a familiar pattern: ownership of large livestock such as cattle or dairy cows typically remains the domain of men, while small livestock such as chickens often belong to and are managed by women. In many African countries, cattle and dairy cows are men's property; cattle ownership bestows respect on men (Quisumbing et al., 2015). The association of women with chickens and men with cattle in Botswana reflects gender/species power relationships (Hovorka, 2012). One current push in agricultural development involves encouraging small farmers and women farmers in particular to be more involved in higher-value agricultural chains including livestock and fisheries. We look at two efforts to promote commercial dairy farming in Africa and one that promotes technological innovation in aquaculture in Bangladesh. In Ethiopia, Lenjiso studies the impact of commercial dairying on men's and women's income and labor. Traditionally milk production in Ethiopia, while low, came under women's control, and women made decisions about whether to consume or sell milk and largely handled the income from the sale of milk. With the introduction of larger-scale dairy farming, new breeds of cattle and new fodder technology, Lenjiso finds control of income becomes defeminized, as men gain more control. At the same time, women's workload increases. Tavenner and Crane examine efforts to promote gender equity in the commercial dairy sector in Kenya. They show how local norms of masculinity related to cattle production complicate the language and tropes of gender equity in the Kenyan intensive dairy sector. Gender equity interventions were framed in the language of patriarchy and/or paternalism and impacted gendered access, participation and benefits from participation in the dairy value chain. Gender norms in Bangladesh aquaculture complete this part of the book. Aregu et al. find that women who used innovative practices and technologies in aquatic agricultural systems earned income and developed a sense of pride as well as recognition in their communities. Evidence suggests that local norms related to gender and aquaculture have begun to shift as women's involvement becomes more socially acceptable.

In most regions of the world, a gender division of labor in agriculture finds women and men performing different types of work within a crop or livestock system. In some cases, men and women raise different crops or livestock, while in other cases they may raise the same crop or animals but for different uses such as subsistence, the local market or the export market. Also, while much of the agriculture in Africa, Asia and Latin America remains small-scale household-level production, the development of export-oriented value chains, especially in fruit, vegetable and flower production, relies heavily on the wage labor of women.

Numerous critiques have been made of the working conditions in the asparagus industry and the reliance on seasonal and temporary workers. Castro Bernardini examines gendered work in the asparagus industry in Peru through an extended case study approach. Using in-depth interviews, ethnographic methods and participant observation, Castro Bernardini found difficult working conditions for most men and women workers in the asparagus fields. Usually, both men and women in households worked in some capacity in the asparagus industry, and working conditions, benefits and wage level varied by gender scale of farm. Looking at trees in agroforestry systems in Nicaragua, Gumucio et al. analyze men's and women's different perceptions and decision-making authority. To develop gender-sensitive, climate-smart agricultural interventions, they collected and analyzed their data. The findings provide a nuanced understanding of different household members' benefits from agroforestry. From this data, we can chart new policies for agroforestry adoption and the promotion of climate mitigation and adaptation efforts.

In Malawi, households with matrilineal and patrilineal land inheritance patterns show gender differentials in agricultural productivity comparing. Bikketi et al. found significant gender gaps in productivity between matrilineal and patrilineal households as well as a growing feminization of agriculture in Malawi. The book concludes with a study of rice value chains in three states in eastern India. Based on a survey of 1,484 farmers, Ganguly et al. find that women's participation in the rice value chain is decreasing except in post-harvesting and marketing. However, women who are involved in the rice value chain are experiencing more participation in decision making.

These scholars collectively reveal the strengths and weaknesses of agricultural development efforts in solving the problems and creating opportunities for men and women engaged in agriculture. They use a range of methodologies as they collect data: surveys, ethnographic methods, participatory action research, focus groups and in-depth interviews. Together, the studies show how, despite the deep entrenchment of practices and norms in agriculture that disadvantage women as farmers, changes are in the works, shifting gender dynamics. Solutions to increasing production, improving the livelihoods of rural people and overcoming food insecurity vary from place to place. Almost always, however, these solutions involve attention to the complicated gender dynamics in agriculture. Context specific and sensitive to how gender intersects with class, caste, ethnicity, age and marital status, these scholars lead the way for future policies and interventions that will empower women and reduce food insecurity.

References

Burke, W., Li, S. and Banda, D. (2018). Female access to fertile land and other inputs in Zambia: Why women get lower yields. *Agriculture and Human Values*, 35(4), pp. 761–775.

Hovorka, A. (2012). Women/chickens vs. men/cattle: Insights on gender/species intersectionality. *Geoforum*, 43(4), pp. 875–884.

Quisumbing, A.R., Rubin, D., Manfre, C., Waithanji, E., van den Bold E., Olney, D., Johnson, N. and Meinzen-Dick, R. (2015). Gender, assets, and market-oriented agriculture: Learning from high-value crop and livestock projects in Africa and Asia. *Agriculture and Human Values*, 32, pp. 705–725.

Udry, C. (1996). Gender, agricultural production, and the theory of the household. *Journal of Political Economy*, 104(5), pp. 1010–1046.

Part 2

Gender and agriculture research

2 The implications of gender relations for modern approaches to crop improvement and plant breeding

Jacqueline A. Ashby and Vivian Polar

Introduction

In many regions of the world, agricultural transformations driven by globalization and the introduction of new technologies perpetuate the disadvantages of women producers compared to men, in terms of access to the key productive resources on which agriculture depends. Food systems – understood as the growing, harvesting, processing, packaging, transporting, marketing, consuming and disposing of food – are shaped throughout by relationships among people with different types and levels of power. Gender inequality occurs in food systems within all types of sociocultural and geographical contexts and at multiple organizational scales, ranging from families to multinational value chains. Gender relations come into play throughout any food system where women and men have different types and levels of control over key resources and decision making about their use – whether in production, trading, employment, agribusiness or food consumption (Rocheleau, 1995; Agarwal, 2001; Fontana, 2009; World Bank, 2012; Quisumbing et al., 2014). FAO calculates that 48% of the economically active women in the world report their primary activity is agriculture, and women make up about 43% of the agricultural labor force globally and from 50 to 70% of the agricultural labor force in Sub-Saharan Africa: thus, gender inequality in agriculture affects more than 2 million people in poor countries, living on almost 500 million small farms (of less than two hectares), who support most of the people living in absolute poverty (FAO, 2011).

Wherever gender defines social norms, roles, rights and responsibilities that justify and support unequal power and control over resources for men and women, they engage in agriculture with different means of production and face different constraints. As a result, women frequently develop different strategies for farming than men, based on systematic differences in their rights, opportunities and resource endowments. In low-income countries where women on small farms produce for direct household consumption as well as for the market, gender inequalities often translate into gender disparities in adoption of new technologies (Peterman et al., 2014; Fisher and Carr, 2015; Mehar, Yamano and Panda, 2017).

Development interventions promoted for overcoming gender gaps in technology adoption frequently emphasize the role of policies designed to level the playing field for rural men and women in terms of their access to land, capital, technology, markets and development services. The equal-access approach assumes that if gender gaps are diminished or disappear, all producers would use the same technology, irrespective of gender (FAO, 2011; Quisumbing et al., 2015). A different approach argues for a diversity of agricultural technologies, some of which may be gender neutral but others of which may be designed to be gender responsive and purposively incorporate a gender bias to benefit women. The disadvantage of this approach is that it may reinforce the underlying gender inequality that creates the need for a gender-specific technology in the first place. The question is whether there is a role for the design of technologies that deliberately target men and women differently without condemning either to relative disadvantage.

Plant breeding provides a useful process with which to explore the case for gender-responsive technology development that involves designing and targeting technology explicitly to benefit women. Gender inequality on the farm is often associated with lower adoption rates of modern plant varieties for women producers compared with men, although the relationship is not straightforward, because sex of the producer is highly correlated with different levels of resource endowment and does not consistently predict a different pattern of adoption (Doss and Morris, 2001; Chirwa, 2005; Smale, 2011; Yiadom-Boakye et al., 2013; Walker et al., 2014; Larson et al., 2015; Letaa et al., 2015; Mutenje et al., 2016; Seymour et al., 2016). For example, a national-scale study conducted in Malawi found gender was an important determinant of modern maize adoption, even after controlling for other individual, household and community characteristics, especially for explaining lower adoption among women in male-headed households (Fisher and Kandiwa, 2014).

The next section provides a brief overview of different stages of a classical plant breeding process to highlight a critical decision point at which breeders could take on board information about gender, when breeding objectives based on a list of prioritized traits is decided. Then, focusing on one type of decision in the breeding cycle, namely trait prioritization, the following section examines what we know about gender differences in plant trait preferences. The remainder of the chapter examines the implications of gender-differentiated trait preferences and important aspects of unequal gender relations in farming for trait prioritization to draw some conclusions about the case for gender-responsive plant breeding.

Plant breeding: trait prioritization and the product profile

For public-sector breeding programs with social welfare and development goals that aim to maximize adoption of their products by resource-poor farmers, a key consideration in product advancement is the likely impact of a new breeding product on the welfare of the end users (farmers) and other beneficiaries in the food system. In the past, modern plant breeding, in particular public-sector breeding concerned with benefitting the poor, has paid very little attention to gender differences in deciding what traits to incorporate in new varieties. This is changing with the need for breeding to become more "client focused" (Ashby, 2009; Persley and Anthony, 2017). As the importance grows of satisfying the client, so the need for breeding to take gender into account is more evident and pressing (Ragot, Bonierbale and Weltzien, 2018). Moreover, with recent technological advances in breeding, such as genomic selection and high throughput processing, opportunities have increased exponentially for plant breeding to expand the range of traits prioritized and tailor new varieties to specific end-user preferences and so address gender-responsive breeding objectives.

A general aim of any plant breeding program is to offer varieties to farmers with advantages for specific traits such as high yield, short growth duration or resistance to an economically damaging pathogen or pest. Improvements of specific traits constitute the basis of successful plant breeding. When breeding programs develop new plant varieties, they usually start with numerous promising candidate materials and progressively reduce them to a select few. This selection process is organized in stages, often called a stage-gate (Kotch, 2018; Ragot, Bonierbale and Weltzien, 2018).

Each stage involves a product advancement decision – that is, a decision on whether to advance a candidate material to the next stage of breeding. The first stage, usually called "product design," is critical for determining outcomes, as this is when information about needed improvements and customer demand is taken into account. This is when breeders decide on the desired plant ideotype or Product Profile. The profile describes, for a well-defined target customer segment, the full set of targeted attributes that a new plant variety is expected to meet to be successfully released onto a market. Attributes are traits with a specific level defined in either absolute or relative terms (Ragot, Bonierbale and Weltzien, 2018, p. 5). The final Product Profile defines, therefore, the package of traits that will be used to set breeding objectives throughout the selection process.

In order to arrive at a final Product Profile, breeders evaluate, weight and prioritize the individual plant traits under consideration for inclusion in the product profile. Trait prioritization is highly selective, because the number of traits that can be included in any one profile is usually restricted to prevent the selection process from becoming unduly complex. The criteria breeders use for trait prioritization are often a mix of commercial, technical and business considerations, shaped by the goals of the breeding program.

Trait prioritization and the definition of the Product Profile is a critical point in the breeding process for taking gender considerations into account. A flawed product design – that is, a variety with unattractive traits from a woman user's perspective – will inexorably lead to low rates of adoption if the woman has any decision-making power in the adoption decision. User acceptance of new varieties in small farm systems of poor countries is heavily dependent on preferences for varietal traits, and men and women often express different trait preferences. There is a practical challenge, therefore: how to systematize relevant information about gender differences, especially men's and women's trait preferences, in a way that breeders can factor it into their trait prioritization and product profiles.

Gender-differentiated plant trait preferences

Product design was identified as a critical stage for consideration of gender in the breeding process, when desired plant traits are prioritized, based on the breeders' product profile that provides a set of breeding objectives. Construction of the product profile involves consideration of end users' criteria for acceptance of the proposed new variety or varieties, and this is when information about gender differences in trait preferences is highly pertinent for breeders' decisions about the product profile. This section examines what is known about gender-differentiated trait preferences.

Explaining the influence of gender on variety adoption rates requires an understanding of gender-differentiated trait preferences (Fisher and Kandiwa, 2014). A recent search of scientific data bases identified only 39 papers that provide evidence, quantitative or qualitative, documenting trait preferences of both men and women, or women only, for a given crop and that also include explanations for those differences (Christinck et al., 2017). The 39 studies are of worldwide origin, with a majority from sub-Saharan Africa; cereals are most represented, followed by legume, root and tuber and other vegetatively propagated crops. The 39 studies consistently report a number of preferred traits that are either important exclusively to women producers or that are given more importance by women producers than by men (Table 2.1). Several of the traits preferred by women, for

Table 2.1 Differences in trait preferences of men and women producers: data from 39 empirical studies

Traits mentioned only by women	*Traits mentioned only by men*
Vigor	Pest resistance
Well adapted to a diversity of growing conditions	Adapted to intercropping
Leafiness	Yield/ha
Storage life	Suitability for local dish
Ease of dehulling	Resistance to water-logging
Ease of threshing	
Quantity of useable flour	
Fuelwood quantity from Stover	
Cooking time	
Taste, grain color	
Tall height for ease of harvest	
Traits mentioned more often or ranked higher by women than by men	
Production-related traits	*Post-harvest traits*
Earliness	Food security
Ease of harvesting and transport	Threshability
Grain traits	Cooking quality
Pest and disease resistance	Less decortication, dehulling, milling losses
Multiple harvests	Market value

Production-related traits	*Post-harvest traits*
Requirements for weeding	Resistance to storage pests Straw quality for roofing Processing quality for locally marketed product Grain and leaf quality Medicinal properties Taste of specific dishes

Source: Christinck et al., 2017, Tables 2.3 & 2.4

example "ease of threshing," are clearly associated with women's customary farm work responsibilities, such as harvesting. Several of the quality traits prioritized by women, such as cooking time, taste and grain quality, reflect their primary responsibility for food preparation, and some of these are also important to men.

A national-scale survey of cassava producers in Nigeria that collected sex-disaggregated data on trait preferences is especially insightful because it shows that generalizable patterns of gender differences in trait preferences for cassava exist at scale (Bentley et al., 2017). For example, taking all study sites together, the study finds that for women, processing traits are crucial, especially ease of peeling of the cassava root, the top preferred trait for women. For the men, agronomic traits were generally more important than processing qualities. Men rarely mentioned ease of peeling, although they did know of some processing traits, such as color, that can affect marketability. The study highlights the difference in traits seen as desirable by spouses (women) compared with heads of households (men, who head 98% of households surveyed). Generally, differences of preferences between spouses (women) and household heads (men) were statistically significant. Spouses (women) valued quality traits more than men did – e.g., poundability, starch content, tapioca taste, ease of load, white color, stem yield, yellow color. Poundability and ease of loading were considered unimportant by men. However, for some other traits such as garri taste, root yield, market demand and storage, there was no significant difference between spouses (women) and household heads (Bentley et al., 2017) Although men and women mostly agreed on the ranking of different varieties, they singled out different traits to explain why they preferred a given variety (Teeken et al., 2018).

Different trait preferences for cassava of men and women producers identified in this study reflect the gendered division of labor that prevails in the Nigerian small-farm cassava production and processing economy, with women, who are primarily responsible for processing, placing a high value on traits that affect their workload or the quality and profitability of their processed cassava foods (Teeken et al., 2018). The different trait preferences also reflect the importance to women of maintaining control over some of the revenue from the crop. For this reason, in some locations, women like cassava varieties that keep well when left in the ground. This trait enables women to harvest a few cassava plants at a time for sale whenever they need money for expenses that they are expected to cover, such as family school fees, clothes, medicines or purchased food (Bentley et al., 2017).

Men and women do not always have different trait preferences, and when they do, these differences are not universal. One reason is that preferences reflect the prevailing division of labor, which varies with the farming system. For example, women responsible for weeding lowland rice in West Africa preferred varieties that suppress weeds more than men (Gridley, 2002) whereas in upland rice in Ghana, men responsible for weeding rice valued weed suppression as a plant trait (Dorward et al., 2007). Moreover, the gendered division of labor is itself subject to change, as when men migrate and so leave many production tasks to women or new market opportunities develop: an example is the feminization of cassava production in southeast Nigeria, which is associated with high levels of empowerment of women cassava producers (Gurung, 2016; Teeken et al., 2018). Gender differences in trait preferences occur when men and women are producing and using crops under very different ecological growing conditions, labor conditions or market conditions – with correspondingly different requirements for plant performance (Hamazakaza et al., 2014). An ecological difference is often the result of underlying gender inequalities, such as women having less access to fertile soil (Dorward et al., 2007; Leiser et al., 2012). Women frequently face different market conditions such as when, for example, gender norms dictate that women do not market grain or that women are only permitted to trade in local markets. In West Africa, women dominate cassava processing and marketing, but this division of labor has different effects on preferences, depending on the markets and resources available to women (Teeken et al., 2018). Different preferences arise from the way labor is organized because gender norms dictate that women supply unpaid farm labor in certain production tasks, such as weeding, and because of the "time poverty" that results from women's heavy burden of reproductive and domestic work on small farms (Blackden and Wodon, 2006).

It is clear that gender-differentiated trait preferences can only be interpreted fully if analyzed in conjunction with information about how rights and responsibilities in the food system are customarily distributed between men and women. The next section discusses how this requirement for analysis of complex gender relations, that vary greatly by culture and region, might be addressed for the purposes of informing breeders' trait prioritization so that these can take gender into account.

Gender screening of plant traits

In plant breeding, the Product Profile is used to define a set of heritable traits that is both biologically feasible to achieve through breeding and will meet demand from a well-identified set of end users or customers. Table 2.2 is an example of a breeders' Product Profile Proposal compiled for prioritizing among a list of 20 traits under consideration for a breeding product. The priority traits that make up the final Product Profile will be the result of comparing the scores in the matrix. In the example in Table 2.2, each of the 20 possible traits is evaluated for market

Table 2.2 Example of a generic Product Profile Proposal with numerous traits requiring prioritization

Trait	*Trait Category*	*Reference*	*Market Priority*	*Selection Objective*	*Desired Short-Term Impact*	*Desired Long-Term Impact*
Grain yield	Agronomy	Check 1 +5%	1	*Maximize*	3	1
Stover yield	Agronomy	Check 5	1	*Maximize*	11	8
Early vigor	Agronomy/ labor	Check 1	2	*Reach threshold*		
Tillering	Agronomy	Check 6	2	*Reach threshold*		
Lodging resistance	Agronomy	< 20%	2	*Reach threshold*	2	3
Plant height	Labor	90–110cm	2	*Reach threshold*	7	5
Maturity	Agronomy	Early	3	*Reach threshold*	9	9
Thousand kernel weight	Agronomy/ seed supply	40–50g	3	*Reach threshold*		
Seed size	Seed supply/ processing	Check 3	2	*Reach threshold*	12	12
Milling yield	Processing	Check 2 + 5%	1	*Maximize*	6	11
Texture	Processing/ consumer	Check 6	3	*Opportunistic*		
Protein content	Consumer	> 7.5%	1	*Maximize*	1	6
Oil content	Consumer	5–8%	3	*Reach threshold*	10	10
Fiber content	Consumer	> = Check 3	2	*Reach threshold*	5	4
Mildew resistance	Agronomy	Resistant	1	*Maximize*	8	7
Virus resistance	Agronomy	> = Check 4	1	*Reach threshold*	4	2
Smut resistance	Agronomy	Intermediate	3	*Opportunistic*		
Heat tolerance	Agronomy	Intermediate	2	*Reach threshold*		
Drought tolerance	Agronomy	Check 5	2	*Opportunistic*		
Al tolerance	Agronomy	Moderate	3	*Opportunistic*		

Source: Adapted from Ragot, Bonierbale and Weltzien, 2018 Table 2.4.

impact, scored as follows: Required = 1; Important = 2; Nice to have = 3. The top 12 traits are also ranked for desired long-term and short-term impact. The practical challenge for gender-responsive breeding is how to integrate gender considerations into this Product Profile Proposal matrix so that each possible trait is also assessed and ranked with respect to its potential acceptability for women as well as for men. The need, therefore, is for a gender screening of possible traits that breeders can include in their Product Profile Proposals.

A framework for gender screening of plant traits

This section proposes a framework for gender screening of plant traits that is intended to be compatible with use of a breeders' Product Profile Proposal. The framework is diagrammed in the form of a Decision Tree in Figure 2.1. This version of the framework is designed to screen a given trait for a given target customer segment, specifically for aspects that are either disadvantageous or uniqussssely advantageous to women. This restricts the analysis to "minimal" effects on women related to gender equity. If the framework is applied to the same plant trait substituting the word "women" for "men" in Figure 2.1, the result will be a complementary analysis related to effects of the plant trait in question on men. Comparison of an assessment of a trait for men and for women is desirable

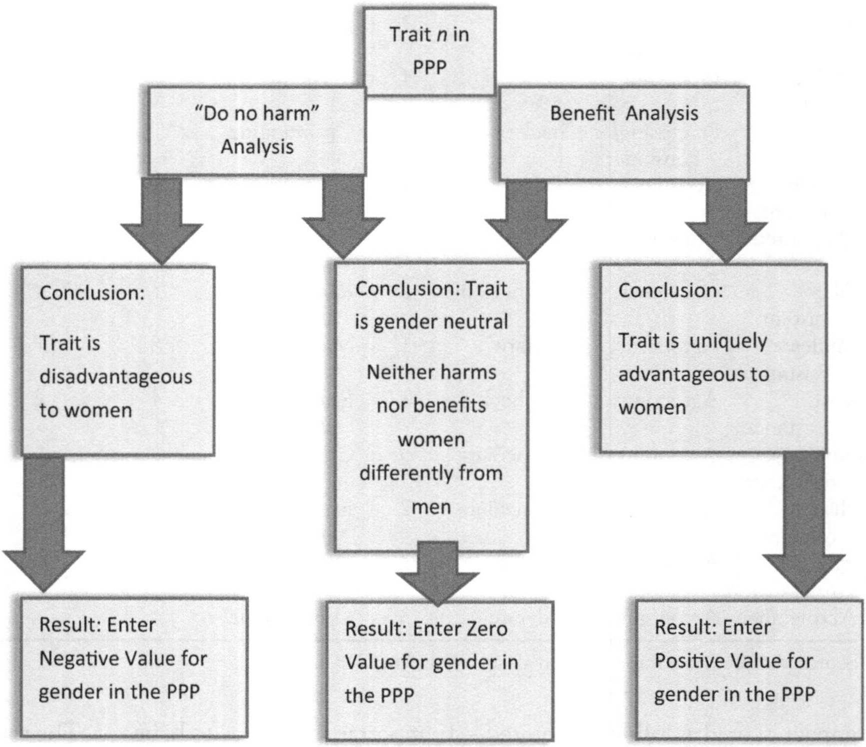

Figure 2.1 Decision tree for gender-screening a trait in a breeding Product Profile Proposal (PPP)

to clarify where there is common ground between men and women producers as well as potential trade-offs involved in targeting men and women producers differently, but limitations of space do not permit a full exploration of this here.

A first consideration is that the final breeding Product Profile needs to include "must-have" traits that should be above a defined threshold for acceptability, since inattention to these will very likely result in non-adoption or dis-adoption of the new variety (Polar et al., 2017; Ragot, Bonierbale and Weltzien, 2018). From a gender perspective, the first objective is to minimize the risk of releasing a variety that could lead to disproportionate non-adoption among women and exacerbate existing gender inequities. For the purposes of a gender screening analysis, this is termed "do no harm" (see Figure 2.1). Traits that do not "pass" the "do no harm" analysis should be qualified as undesirable in the Product Profile Proposal from a gender perspective. A negative result of a gender screening is a signal that gender-blind prioritization of traits risks overlooking a potentially harmful outcome (see Do No Harm in Figure 2.1).

An important proviso is that breeding is not necessarily the only or best way to overcome a constraint expressed in a trait preference: for instance, a more cost-effective way to address the preference for "easy to peel" cassava might be to introduce affordable, small-scale mechanized peelers. The benefit of including "do no harm" gender screening into trait prioritization, using the Product Profile Proposal, is that discussion of how best to address gender-related constraints can take place when a breeding product is designed and is not left until much later, when low rates of adoption signal there is a problem.

Traits that "pass" the "do no harm" gender screening can be qualified as either gender neutral (neither beneficial nor harmful to women differently from men) or as conferring a positive advantage or benefit specifically targeted at women (see Figure 2.1). This outcome illustrates that the result of this gender screening is not to require a final Product Profile to include traits that are preferred by women. After completing gender screening, a final Product Profile may end up by prioritizing traits that are all gender neutral. The requirement that all traits undergo a gender screening reduces the risk that gender-blind trait prioritization has overlooked the possibility of a harmful outcome. In addition, the advantage is that beneficial aspects of traits in the proposed Product Profile can be identified, and this information can be used to weight traits and assess trade-offs between alternative traits when they are prioritized. Use of gender screening should enable discussion of how much importance a breeding project wants to give to designing a gender-responsive product by prioritizing traits that have benefits for women or by adding a new trait with benefits for women to the Profile. Use of gender screening creates the opportunity for a breeding project to opt for including a highly beneficial, "game-changing" trait in design of a new breeding product that is potentially transformative.

A gender-responsive breeding program should at least opt for a value of zero (neutral outcome in Figure 2.1) with respect to "do no harm" for traits in a final Product Profile. A breeding program that aims to positively benefit women will opt for a positive result from gender screening for one or more traits included in the final Product Profile (see Benefit Analysis in Figure 2.1). Trait prioritization involves assessing trade-offs between alternative outcomes, and gender screening can help to make trade-offs explicit between different preferences. Breeders can then decide, for example, whether gender-responsive breeding should opt for improvement of "niche traits" or "game-changing" traits (Mascarenhas, 2016; Kotch, 2018). Niche traits, such as specific qualities required for food processing, are often strongly preferred by women and can address their practical needs by improving the profitability of women's income generation or by relieving their drudgery. "Game-changing" traits, such as "earliness," can be transformative by altering the design of a proposed variety in order to overcome a critical obstacle identified by the gender screening that specifically disadvantages women producers (such as late planting due to chronic delays in land preparation) or to optimize a desired benefit of especial benefit to women producers (such as shortening the "hungry season").

Criteria for gender screening of plant traits

This section discusses criteria that are essential for gender screening of plant traits for prioritization in breeding. These criteria are not the only ones that could be applied for the purpose of trait gender screening, but they represent a minimal "must-have." The criteria are drawn from aspects of gender inequality that stand out in the expressed plant traits preferred by women discussed earlier. In Box 2.1, each criterion is expressed as a question to screen for a trait's negative aspects in Do No Harm Analysis. In Box 2.2, Benefit Analysis uses questions to screen for a trait's positive aspects based on four of the five criteria.[1]

Effect of the proposed trait on women's unpaid labor

The review of trait preferences shows the value women give to traits that alleviate (or at a minimum, do not increase) the burden of unpaid work and drudgery they do on small farms: i.e., their "time poverty" (Blackden and Wodon, 2006). Time poverty refers to the situation in which an individual does not have enough time for rest and leisure after working, whether in the labor market, for domestic work, or for other activities such as fetching water and wood. A general finding is that women do considerably more unpaid farm labor than men. Women's average daily hours in agricultural work in four Sub-Saharan African countries is almost 467 minutes a day, compared with about 371 minutes a day for men (Blackden and Wodon, 2006). A plant trait can increase the burden of work on small farms and decrease the returns to women's unpaid labor if it impacts the amount of time that the family, and women in particular, devote to production, harvesting or processing. For example, women in Ethiopia objected to modern short-strawed sorghum varieties that would increase their workload (Mulatu and Belete, 2001). In East Africa, maize adoption lagged because women objected to hard-dent maize varieties that were difficult to grind and so increased their workload (Jha, Hojjati and Vosti, 1991).In Uganda, about 78% of hybrid banana varieties evaluated for release had low heat-retaining capacity, which made the cooked food harden very fast when served. The hybrids required prolonged cooking time to soften the texture and so required more labor time to gather fuelwood, which represented an additional burden to the already overworked women (Sanya et al., 2018).

On the other hand, plant traits can have the benefit of alleviating the demand for women's unpaid labor, as several of the traits identified in Table 2.1 illustrate. Women preferred tall rice varieties for ease of harvesting (Gridley, 2002; Manzanilla et al., 2014); long pearl millet panicles for easier handling (Baidu-Forson, 1997); traits conferring ease of threshing for sorghum and rice (Kudadjie, 2006), pearl millet (Baidu-Forson, 1997); (Dorward et al., 2007; Manzanilla et al., 2014; Isaacs et al., 2018). In Nigeria, women farmers in all regions surveyed preferred early-maturing cassava varieties to reduce the number of times that they need to weed (Bentley et al., 2017).

Effect of the proposed trait on activities for own income generation

Time poverty is a crucial aspect of gender inequality for consideration in plant breeding because plant traits that increase the amount of unpaid work women do in that crop can displace their other income-generating activities, including paid work on other farms. If a new variety increases workload, this can mean they have to take time away from other productive activities, such as growing or processing food for consumption or sale, and consequently experience a loss of income under their own control. The especial emphasis here is the effect of a trait leading to displacement of women's activities that contribute to income under their own control, because many of the trait preferences reviewed express this concern for use of plant resources for this purpose: control over income from their labor is a crucial aspect of well-being for the individual woman and for the welfare of children who benefit from maternal income used for food, schooling and medicines (Alkire et al., 2013). In Nigeria, when cassava is difficult to peel, women have less time available for other important tasks and end up cutting off some of the good root with the peel. Being hard to peel not only increases women's labor but also lowers the income they can generate from the crop (Bentley et al., 2017). Women in Ethiopia objected to modern short-strawed sorghum varieties that meant they lost income from sale of the stover for cooking fuel (Mulatu and Belete, 2001). In West Africa, women were critical of NERICA rice because it increased their field labor in weeding and bird scaring (Lodin, Paulson and Mugenyi, 2012; Lodin, 2012). The extra time devoted to weeding and bird scaring displaced time otherwise devoted to other food crops (Lodin, 2012).

Numerous plant traits are beneficial and valued by women because they are important for own income-generation activities or increase returns to women's labor. In Malawi, women who sell leaf sauce in the local market value cowpea and cassava varieties with suitably edible leaves (Chiwona-Karltun et al., 1998; Kitch et al., 1998). Similarly, in Nigeria, women who processed cassava foods for income generation (*gari, fufu and Abacha*) prioritized traits important for these products: sweetness, low in fiber, low in moisture, easy to peel, suitable food color (cream when toasted into *gari* and white when processed into *fufu and Abacha*) (Bentley et al., 2017).

Effect of the proposed trait on control over production inputs

It is common for performance of modern plant varieties to incorporate a trait that depends critically on access to an essential production input (e.g., water, land, fertile land, labor, capital, transportation, knowledge, technology, weed killer, pesticides, fungicides) that women have greater difficulty obtaining than men. If a new plant trait changes the producer's requirement for a complementary input, such as a new variety that relies on application of chemical fertilizer to outperform the current varieties, and women customarily struggle to access this input or have to reallocate their own income away from school fees and medicine to buy it, then the trait is potentially disadvantageous to women. In Tanzania, the introduction of new forage systems depended on women having enough power and status in the community to claim use of two crucial inputs – land to establish a traditional grazing area or *ololili* and the resources for building fences to protect the *ololili*: both were difficult for women livestock keepers, particularly for poorer women and widows. Forage intensification with drought-resistant crops could cause more invasions by neighbors stricken by extreme drought. This would mean that, as a result of the intervention, women farmers who could not defend the *ololili* would lose all their forage and thereby be worse off (Galie et al., 2018).

Effect of the proposed trait on control over production outputs

The impact of a new plant trait on women's control of products and by-products can be insidious, simply because of the knock-on effects when increased production from use of a modern variety generates increased surplus for sale. It is difficult to classify certain crops definitively as "women's crops" because women contribute to the production of most crops either on plots they control, on plots men control or on plots they control jointly (Doss, 2002). What matters here is control over the product, especially as a source of food, and control over the revenue from sales. In several African countries, as total income generated from the commodity increased over time, there was a decline in women's control of income in five different commodities, including beans, groundnut and soybeans (Njuki, Kruger and Starr, 2013). Women in most cases control the production of food crops for household consumption and any income that arises from these crops. Women also tend to control the income derived from semi-subsistence crops (Njuki et al., 2011) With commercialization, there is loss of control of the products of their labor, and when men take over production, more of the food may be sold and the revenue is not automatically shared back with the women who did the work to produce it. They therefore have less income under their direct control for food, health and other household essentials (Paolisso and Regmi, 1992; Dolan and Sorby, 2003; Fischer and Qaim, 2012; Rubin and Manfre, 2014; Seguino and Were, 2014; Johnson et al., 2016) Dolan (2001) documents how the commercialization of French bean production in Kenya, a traditional domain of women, has been appropriated by men. In another example, banana has traditionally been a women's semi-subsistence crop in Kenya but is gaining importance as a cash crop. When several thousand banana producers were organized in groups to facilitate access to disease-free planting material, improved varieties, credit and markets, women producers tended to lose control over the produce and revenues to men (Fischer and Qaim, 2012). Women who produce NERICA rice in Uganda complained that "Even if the woman is planting the rice, the man is selling it." This was prevalent when the spouses produced rice on a joint household plot, but even when the product came from the wife's own plot, it was usual for the husband to market the grain and then give some portion of the revenue to the woman. Men had the final say on the quantity of rice that would be sold and usually insisted on selling off a major part of the produce, even if their spouses wanted to keep some for cooking (Lodin, 2012; Yokouchi and Saito, 2016).

Plant breeding based on the premise that increased productivity is a "must have" for new varieties intended to benefit the poor and hungry cannot always directly influence the loss of control resulting from increased commercialization of a crop, discussed above. However, in breeding with a client-focus, the importance is growing for breeding to incorporate quality traits such as those expressed in Table 2.1. There are many instances in which women producers value a quality trait over and above increased yield, because local varieties involve this trade-off. In Mali, sorghum grain produced by women on their own plots is vital for feeding children outside the main meals and when the men's granaries run empty. Women prefer early-maturing sorghum varieties suitable for cultivating as an intercrop in their groundnut fields. In contrast, men, who are responsible for producing sorghum as the staple for the main family meals, prefer higher-yielding varieties for cultivating as sole crops (Isaacs et al., 2018). Gender screening of individual plant traits will highlight that there is a gender dimension of choices to be made between optimizing yield or a different, desirable trait. By including gender screening of traits in trait prioritization, the advantages and disadvantages of prioritizing a preferred trait over and above increased yield can be systematically weighed, taking the gender implications into account.

Gender-differentiated trait preferences

Gender-differentiated trait preferences provide crucial information for prioritizing traits in breeding, but they cannot always be taken at face value and need to be complemented with the above analysis of gender inequality for several reasons. When men and women agree on a negative evaluation of the trait in question, this is a signal that adoption may be limited and that breeders cannot afford to ignore. When men and women agree on a positive evaluation of a trait, this is a compelling expression of consumer demand and a signal that the trait should be retained in the new variety. However, the reasons for preferences should always be considered: men and women may value the same trait equally but for different reasons that relate to their division of labor, rights and responsibilities in relation to the crop. Those reasons are usually tied to differing objectives for the crop that can have implications for trait prioritization. As the review of gender-differentiated trait preferences identified, there may be disagreement about the relative value of the same trait among men and women in the target customer segment. Divergent perceptions of the same trait are a clue to the potential existence of conflicts of interest that reflect underlying gender inequality (for example, when the majority of men are positive about a trait that confers an advantage in the market, but the majority of women are negative, perceiving that the trait increases their workload). Another example is when women value a trait for its food value or cooking quality while men discount this in favor of marketability, as seen in the Nigeria cassava study. In Uganda, there was a significant difference between women and men in the overall acceptability score for hybrid M9 banana: women who cooked the banana appreciated the food value of hybrid M9 more than men did. Men expressed low acceptance due to the failure of M9, despite the big bunch, to attract a higher market price (Sanya, 2018). In central Malawi (Chiwona-Karltun et al., 1998) unlike the men, the poorer women reject sweet for bitter types of cassava because bitter cassava is less likely to be stolen from their fields, despite the extra post-harvest labor required to remove the toxic compounds from bitter varieties. In Tanzania, women preferred risk averting traits of new finger millet varieties, like short duration, drought tolerance, compact heads and disease resistance, while men preferred different, market related traits like high yield, brown color and large heads (Ojulong, 2017).

Conflicts of interest may be resolved by bargaining over what to plant: men or women may or may not exercise their preference in an adoption decision. The analysis of gender relations should provide insight on whose opinion is likely to predominate in bargaining and why. When trait preferences of men and women producers conflict, breeders' trait prioritization may involve deciding which members of the customer segment to favor or to disoblige. Disagreement about the value of a trait among men and women in the target customer segment, or among women who also may have divergent interests in plant performance, can suggest the need to refine the classification of the "customer" for a product if there are subgroups of women or men with different preferences.

Application of gender screening of plant traits

The objective of the gender screening proposed is to include a value (positive, neutral or negative) in the product Profile Proposal used by breeding programs to evaluate and then prioritize traits being considered for inclusion in the final definition of the breeding product or variety. In addition, gender screening may identify new desirable traits from a gender perspective, for assessment by breeders and that may deserve inclusion in the Product Profile. To date, the framework for gender screening and method for its application have been reviewed by a multidisciplinary group of breeders and social scientists, who applied it to several cases documenting gender-differentiated trait preferences in a workshop preparatory to its field testing. They successfully generated a discussion of the gender implications of individual plant traits along with values that they considered could be part of a breeding program's product advancement discussion and overall product profile design process (GBI, 2018).

For this purpose, the criteria for gender screening discussed in the previous section were used to formulate six questions for a "Do no harm" analysis (Box 2.1) and five questions for a "Benefit Analysis" (Box 2.2) used for gender screening, following the Decision Tree for Gender Screening of trait (Figure 2.1). Answering these questions requires informed judgments based on detailed evidence about gender differences in the target customer segment that can be generalized with confidence. A scoring procedure was provided so that a positive, neutral or negative value could be generated from the answers to the questions (GBI, 2018).

Box 2.1 Gender Screening of Plant Traits

Do No Harm Analysis: for each trait in the Proposed Product Profile

Question 1. Drudgery: Does the trait involve a harmful increase in the unpaid, family labor input by women in the target customer segment, to production or to use of the product, including marketing or processing for household consumption or sale?

Increases unpaid labor with serious harm to individual women's and/or household's health and welfare for a majority in the target customer segment = −2
Increases unpaid labor with moderate harm for a majority OR variety release is conditional on implementation of a guaranteed remedial intervention to remove this obstacle for a majority = −1
No significant increase = 0
Warning: not enough information to score: please note in the PPP

Question 2. Displacement of women's productive activity: Can the trait involve a reduction in, or removal of a major activity for production, use or marketing that women in the target customer segment rely on for income generation?

Displaces women's activities with serious loss for a majority of women in the target customer segment = −2
Displaces women's activities with moderate loss for a majority = −1
No significant displacement for a majority = 0
Warning: not enough information to score: please note in the PPP

Question 3. Control of production inputs: Can crop improvement associated with the trait depend critically on access to an essential input (e.g., water, land, fertile land, labor, capital, transportation, knowledge, technology) that women in the target customer segment have greater difficulty obtaining than men?

Access to and control over least one essential input is seriously unequal for a majority of women in the target customer segment = −2
Access to and control over least one essential input is seriously unequal for a majority of women OR variety release is conditional on implementation of a guaranteed remedial intervention to remove this obstacle for a majority = −1
No prevailing gender inequality in access to an essential input for a majority = 0
Warning: not enough information to score: please note in the PPP

Question 4. Control of products and by-products: Can the trait remove or reduce the quantity or quality of a product (e.g., a traditional intercrop) or by-product (e.g., fodder), a source of food (e.g., edible leaves) or source of income generation normally under the control of women in the target customer segment?

Loss of product or income from the product will be incurred with serious harm to health and welfare of a majority of women and/or their children in the target customer segment = −2
Loss will be incurred with moderate harm to health and welfare of a majority of women and/ or their children = −1
No loss for a majority = 0
Warning: not enough information to score: please note in the PPP

Question 5. Negative Trait Preferences: Do women in the target customer segment value the trait negatively?

Yes: the majority value it negatively = −1
NO: most women do not value the trait negatively = 0

Question 6. Negative Trait Preferences: Do men in the target customer segment value the trait negatively?

Yes: the majority value it negatively = – 1
NO: Most men DO NOT value the trait negatively = 0

An example of gender screening of a trait using the questions in Boxes 2.1 and 22.2 is based on a case study of a target customer segment for cassava breeding who comprise small-scale men and women cassava producers in southwest Nigeria (Olaosebikan et al., 2018). The trait for screening was “increased shelf life of fresh cassava roots.” With respect to the Do No Harm analysis, the screening for drudgery concluded that improved shelf life can increase the return to

Box 2.2 Gender Screening of Plant Traits

Benefit Analysis: for each trait in the Proposed Product Profile

Question 7. Reduce drudgery.
Can the trait involve a beneficial decrease in the unpaid, family labor input to production or to use of the product by women in the target customer segment, including marketing or processing for household consumption or sale?

- Decreases drudgery with significant improvement for a majority of individual women's and/or household's health and welfare = +2
- Decreases drudgery with moderate benefit for a majority = +1
- No significant decrease for a majority = 0
- Not enough information to score; Please note in the PPP

Question 8: Employment for own income generation.
Can the trait maintain or increase employment, as hired field labor on or off farm, or in agroenterprise, that women in the target customer segment rely on for generating income under their discretionary control?

- Increases or maintains women's employment with significant gain in own income for a majority of women in the target customer segment = +2
- Increases or maintains women's employment with moderate gain in own income for a majority = +1
- No significant change for a majority = 0
- Not enough information to score; Please note in the PPP

Question 9: Are women-controlled products or by-products increased and/or improved?
Can the trait maintain or increase the quantity or quality of an important product or by-product of the variety or animal (e.g., food, seed, fodder, manure, fuel) normally under women's control?

- At least one important product or by-product under the control of women is significantly improved for a majority of women in the target customer segment = +2
- At least one important product or by-product under the control of women is moderately improved for a majority = +1
- No product or by-product under the control of women is affected for a majority = 0
- Not enough information to score; Please note in the PPP

Question 10. Do women in the target customer segment value the trait positively?

Yes: the majority of women value it positively = +1
No: most women do not value the trait positively = 0

Question 11. Do men in the target customer segment value the trait positively?

Yes: the majority of men value it positively = +1
No: most men do not value the trait positively = 0

women's labor in processing fresh roots, and this may benefit women. In relation to the second question, whether the trait has implications for displacing women's own income-generating activity, improved shelf life could lead to higher prices of the raw material for processing foods for sale on which women depend for own income, thus reducing their profits. On the third question of whether the trait relies on a critical input to which women have unequal access, market access was considered relevant because the market for cassava is segmented by gender. The majority of women producers are growing and buying roots as raw material for their processing activities, to sell food locally. Men are mainly interested in sale of fresh roots to distant markets. Longer shelf life may increase prices and enable access to a more profitable market for fresh roots sold by men, but this competition is likely to reduce the availability and affordability of fresh roots for women processors. With respect to the fourth question, the effect of increased shelf life on women's control over product, it will enable a more profitable market for fresh cassava, but it might deprive the women of one of the opportunities they have to make money used for their own discretionary spending (food, child welfare). With respect to questions five and six on trait preferences, the analysis found men and women diverged: men valued yield and market qualities, while women's top-ranked traits referred to other processing qualities. After scoring, the gender screening generated a negative value on the Do No Harm analysis. More importantly, it provokes the question of whether improved shelf life is more deserving of inclusion in the Product Profile than other traits valued by women that would improve their local income generation.

Conclusions

One of the persistent gender gaps in agriculture is lower adoption rates of modern varieties among women producers, although gender is by no means a universal or consistent predictor of adoption. Studies of gender-differentiated preferences for plant traits throw some light on this picture, showing how varietal choice is related to the different ways in which resources, rights and responsibilities are shared among women and men engaged in small-farm production, processing and marketing. Plant breeding has had difficulty taking gender differences into account when plant ideotypes, product design and breeding objectives are decided, in part because of the complexity of this picture. Added to this, much of the published information is inadequate for this task: it consists of a description of a trait preferred by women or ranked higher by women than by men, for example "earliness," without an explanation of the desired extent or level of the trait. This limits the usefulness of the information to breeders, who need to understand what producers consider the desired performance level of a trait. Trait preferences are also reported without analysis of the socioeconomic characteristics of respondents other than their gender and geographic location. Simple sex-disaggregation of preference data is not very useful for informing breeding objectives, because it is essential to understand what resource constraints or producers' objectives are associated with a given preference and whether there is an underlying gender inequality at work. In addition, data on gender differences in trait preference studies is too often reported without evidence that the respondents are representative of a clearly identified population of end users. This makes it difficult to draw general conclusions about the significance of a gender-differentiated trait preference at a scale that a breeding program can rely on, as predictive of widespread end-user acceptance.

Nonetheless, the decision made by breeders to prioritize traits for a final product profile is a key decision point when information about gender can be introduced into the design of new varieties. This chapter has laid out the rationale for focusing on the gender screening of individual plant traits in a way that sets the stage for an evidence-based discussion between breeders and experts in gender analysis about what gender means for trait prioritization and for broader breeding objectives that aim to be gender responsive. The review of gender-differentiated traits highlights that women as well as men value plant traits that are adaptive for their resources and constraints. Gender screening of plant traits does not produce a case for the development of specifically "women's" varieties, but it does inform key breeding decisions to make it more likely that breeding will include traits important for women.

Acknowledgments

The research on which this Chapter draws was supported by the Gender and Breeding Initiative www.rtb.cgiar.org/gender-breeding-initiative/, coordinated by the CGIAR Research program on Roots, Tubers and Bananas www.rtb.cgiar.org, and with support in 2016–2017 from the CGIAR System Office Gender Research and its CGIAR donors. All opinions are the sole responsibility of the authors.

Note

1 In the questions for Benefit Analysis, it is theoretically possible to screen for whether there are positive aspects of a trait that increase returns to women from a production

input over which women have special access or control. An example would be special knowledge of a plant's medicinal or nutritional qualities that is traditionally women's domain. However, what matters is whether women capture such benefits, covered in question nine in Box 2.2.

References

Agarwal, B. (2001). Participatory exclusions, community forestry, and gender: An analysis for South Asia and a conceptual framework. *World Development*, p. 29.

Alkire, S., Meinzen-Dick, R., Peterman, A., Quisumbing, A., Seymour, G. and Vaz, A. (2013). The women's empowerment in agriculture index. *World Development*, 52, pp. 71–91.

Ashby, J.A. (2009). The impact of participatory plant breeding. *Plant Breeding and Farmer Participation*, pp. 649–671.

Baidu-Forson, J. (1997). On-station farmer participatory varietal evaluation: A strategy for client-oriented breeding. *Experimental Agriculture*, 33(1), pp. 43–50.

Bentley, J., Olanrewaju, A., Madu, T., Olaosebikan, O., Abdoulaye, T., Assfaw Wossen, T., Manyong, V., Kulakow, P., Ayedun, B., Ojide, M. and Girma Tessema, G. (2017). *Cassava farmers' preferences for varieties and seed dissemination system in Nigeria: Gender and regional perspectives*. IITA Monograph, Ibadan: IITA.

Blackden, C.M. and Wodon, Q. (eds.) (2006). *Gender, time use, and poverty in Sub-Saharan Africa. World Bank Working Paper; No. 73*. Washington, DC: The World Bank.

Chirwa, E.W. (2005). Adoption of fertiliser and hybrid seeds by smallholder maize farmers in Southern Malawi. *Development Southern Africa*, 22(1), pp. 1–12.

Chiwona-Karltun, L., Mkumbira, J., Saka, J., Bovin, M., Mahunguy, N.M. and Rosling, H. (1998). The importance of being bitter – A qualitative study on cassava cultivar preferences in Malawi. *Ecology of Food and Nutrition*, 37(3), pp. 219–245.

Christinck, A., Weltzien, E., Rattunde, F. and Ashby, J. (2017). *Gender differentiation of farmer preferences for varietal traits in crop improvement: Evidence and issues*. Working Paper No. 2. CGIAR Gender and Agriculture Research Network; CGIAR System Management Office and International Center for Tropical Agriculture (CIAT), Cali, Colombia, p. 38. Available at: https://hdl.handle.net/10947/4660.

Dolan, C. (2001). The good 'wife': Struggles over resources in the Kenyan horticulture sector. *Journal of Development Studies*, 37(3), pp. 39–70.

Dolan, C. and Sorby, K. (2003). *Gender and employment in high-value agriculture industries*. Agriculture and Rural Development Working Paper, 7.

Dorward, P., Craufurd, P., Marfo, K., Dogbe, W. and Bam, R. (2007). Improving participatory varietal selection processes: Participatory varietal selection and the role of informal seed diffusion mechanisms for upland rice in Ghana. *Euphytica*, 155, pp. 315–327. doi:10.1007/s10681-006-9333-y.

Doss, C.R. (2002). Men's crops? Women's crops? The gender patterns of cropping in Ghana. *World Development*, 30(11), pp. 1987–2000.

Doss, C.R. and Morris, M.L. (2001). How does gender affect the adoption of agricultural innovations? The case of improved maize technology in Ghana. *Agricultural Economics*, 25(1), pp. 27–39.

FAO. (2011). *The state of food and agriculture. Women in agriculture: Closing the gender gap for development*. Rome: FAO.

Fischer, E. and Qaim, M. (2012). Gender, agricultural commercialization, and collective action in Kenya. *Food Security*, 4(3), pp. 441–453.

Fisher, M. and Carr, E.R. (2015). The influence of gendered roles and responsibilities on the adoption of technologies that mitigate drought risk: The case of drought-tolerant maize seed in Eastern Uganda. *Global Environmental Change*, 35, pp. 82–92.

Fisher, M. and Kandiwa, V. (2014). Can agricultural input subsidies reduce the gender gap in modern maize adoption? Evidence from Malawi. *Food Policy*, 45(C), pp. 101–111.

Fontana, M. (2009). The gender effects of trade liberalization in developing countries: A review of the literature. *Rafael E. de Hoyos y Maurizio Bussolo. Gender Aspects of the Trade and Poverty Nexus. A Macro-Micro Approach*, pp. 25–50.

Galie, A., Kizima, J., Mangesho, W. and Lukuyo, B. (2018). Gender-responsive forage intensification in the ololili system of Tanzania. In: H.A. Tufan, S. Grando and C. Meola, eds., *State of the knowledge for gender in breeding: Case studies for practitioners. Lima (Peru)*. CGIAR Gender and Breeding Initiative. Working Paper. No. 3, pp. 44–54. Available at: https://cgspace.cgiar.org/handle/10568/92819.

Gender and Breeding Initiative (GBI). (2018). *The right tools for the job: Enabling breeding programs to be gender responsive*. Available at: www.rtb.cgiar.org/gender-breeding-initiative/news-and-opinions/tools-enabling-breeding-programs-gender-responsive/.

Gridley, H. (2002). *Participatory varietal selection in West and Central Africa. Quantitative analysis of data from participatory methods in plant breeding*. Mexico: CIMMYT.

Gurung, K., Bhandari, H. and Paris, T. (2016). Transformation from rice farming to commercial aquaculture in Bangladesh: Implications for gender, food security, and livelihood. *Gender, Technology and Development*, 20(1), pp. 49–80.

Hamazakaza, P., Katungi, E., Ryes, B., Maredia, M., Muimui, K. and Ojara, M. (2014). *Assessing access and adoption of common bean improved varieties in Zambia*. Zambia: PABRA.

Isaacs, K., Weltzien, E., Diallo, C., Sidibe, M., Diallo, B. and Rattunde, F. (2018). Farmer engagement in culinary testing and grain-quality evaluations provides crucial information for sorghum breeding strategies in Mali. In: H.A. Tufan, S. Grando and C. Meola, eds., *State of the knowledge for gender in breeding: Case studies for practitioners. Lima (Peru)*. CGIAR gender and breeding initiative. Working Paper. No. 3. Available at: https://cgspace.cgiar.org/handle/10568/92819.

Jha, D., Hojjati, B. and Vosti, S. (1991). The use of improved agricultural technology in Eastern Province. Adopting improved farm technology: A study of smallholder farmers in Eastern Province, Zambia. In: R.T. Celis, J.T. Milmo and S. Wanmali, eds., *Adopting improved farm technology: A study of smallholder farmers in Eastern Province, Zambia*. Washington, DC: IFPRI, pp. 173–201. Available at: https://books.google.co.uk/books?hl=en&lr=&id=BMaFFv0vDWwC&oi=fnd&pg=PA173&dq=agricultural+technology+Jha&ots=1ZnXXr3J1N&sig=zYyKaB3OtiTKhAXBAnz8Oq5yXm0#v=onepage&q=agricultural%20technology%20Jha&f=false.

Johnson, N.L., Kovarik, C., Meinzen-Dick, R., Njuki, J. and Quisumbing, A. (2016). Gender, assets, and agricultural development: Lessons from eight projects. *World Development*, 83, pp. 295–311.

Kitch, L.W., Boukar, O., Endondo, C. and Murdock, L.L. (1998). Farmer acceptability criteria in breeding cowpea. *Experimental Agriculture*, 34, pp. 475–486.

Kotch, G. (2018). *Overview of excellence in breeding platform*. Powerpoint Presentation, July 8, Gender and Breeding Initiative.

Kudadjie, C.Y. (2006). *Integrating science with farmer knowledge: Sorgum diversity management in North-East Ghana*. Wageningen: Wageningen University.

Larson, D.F., Savastano, S., Murray, S. and Palacios-López, A. (2015). *Are women less productive farmers? How markets and risk affect fertilizer use, productivity, and measured gender effects in Uganda.* S.I.:s.n. Available at: http://elibrary.worldbank.org/doi/book/10.1596/1813-9450-7241.

Leiser, W.L., Rattunde, H.F.W., Piepho, H-P., Weltzien, E., Diallo, A., Melchinger, A.E., Parzies, H.K. and Haussmann, B.I.G. (2012). Selection strategy for sorghum targeting phosphorus-limited environments in West Africa: Analysis of multi-environment experiments. *Crop Science*, 52(6), pp. 2517–2527. doi:10.2135/cropsci2012.02.0139.

Letaa, E., Kabungo, C., Katungi, E., Ojara, M. and Ndunguru, A. (2015). Farm level adoption and spatial diffusion of improved common bean varieties in Southern highlands of Tanzania. *African Crop Science Journal*, 23(3), pp. 261–277.

Lodin, J.B. (2012). Intrahousehold bargaining and distributional outcomes regarding NERICA upland rice proceeds in Hoima district, Uganda. *Gender, Technology and Development*, 16(3), pp. 253–278.

Lodin, J.B., Paulson, S. and Mugenyi, M.S. (2012). New seeds, gender norms and labor dynamics in Hoima District, Uganda. *Journal of Eastern African Studies*, 6(3), pp. 405–422.

Manzanilla, D.O., Paris, T.R., Tatlonghari, G.T. and Tobias, A.M. (2014). Social and gender perspectives in rice breeding for submergence tolerance in Southeast Asia. *Experimental Agriculture*, 50(2), pp. 191–215.

Mascarenhas, M. (2016). *The "must-have" features of gender-responsive plant or animal breeding.* Workshop summary. CGIAR Gender and Agriculture Research Network. Cali, Colombia, p. 4. Available at: https://cgspace.cgiar.org/handle/10568/78078.

Mehar, M., Yamano, T. and Panda, A. (2017). Role of gender, risk and time preference in influencing rice variety selection decision in Eastern India. *Asian Journal of Agricultural Development*, 17(1), pp. 17–36.

Mulatu, E. and Belete, K. (2001). Participatory varietal selection in lowland sorghum in Eastern Ethiopia: Impact on adoption and genetic diversity. *Experimental Agriculture*, 37(2), pp. 211–229.

Mutenje, M., Kankwamba, H., Mangisonib, J. and Kassie, M. (2016). Agricultural innovations and food security in Malawi: Gender dynamics, institutions and market implications. *Technological Forecasting and Social Change*, 103, pp. 240–248.

Njuki, J., Kaaria, S., Chamunorwa, A. and Chiuri, W. (2011). Linking smallholder farmers to markets, gender and intra-household dynamics: Does the choice of commodity matter? *The European Journal of Development Research*, 23(3), pp. 426–443.

Njuki, J., Kruger, E. and Starr, L. (2013). *Increasing the productivity and empowerment of women smallholder farmers: Results of a baseline assessment from six countries in Africa and Asia. CARE international.* Atlanta: CARE.

Ojulong, H., Letayo, E., Sakwera, L., Mgonja, F., Sheunda, P., Kibuka, J., Otwani, D., Audi, P., Mgonja, M. and Manyasa, E.O. (2017). Participatory variety selection for enhanced promotion and adoption of improved finger millet varieties: A case for Singida and Iramba Districts in Central Tanzania. *African Journal of Rural Development*, 2(1), pp. 77–93.

Olaosebikan, O., Kulakow, P., Tufan, H., Madu, T., Egesi, C. and Teeken, B. (2018). A case study of cassava trait preferences of men and women farmers in Nigeria: Implications for gender-responsive cassava variety development. In: H.A. Tufan, S. Grando and C. Meola, eds., *State of the knowledge for gender in breeding: Case studies for practitioners. Lima (Peru)*. CGIAR gender and breeding initiative. Working Paper. No. 3, pp. 35–43. Available at: https://cgspace.cgiar.org/handle/10568/92819.

Paolisso, M. and Regmi, S.C. (1992). *Gender and the commercialization of subsistence agriculture in Nepal.* Washington, DC: International Center for Research on Women (ICRW).

Persley, G. and Anthony, V. (eds.) (2017). *The business of plant breeding: Market – led approaches to new variety design in Africa*. Oxon, UK: CABI Wallingford. Available at: www.cabi.org.

Peterman, A., Behrman, J. and Quisumbing, A. (2014). A review of empirical evidence on gender differences in nonland agricultural inputs, technology, and services in developing countries. In: *Gender in agriculture*. Dordrecht: Springer, pp. 145–186.

Polar, V., Babini, C., Flores, P., Velasco, C. and Fonseca, C. (2017). *Technology is not gender neutral: Factors that influence the potential adoption of agricultural technology by men and women*. CGIAR Research Program on Roots, Tubers and Bananas. La Paz (Bolivia). International Potato Center, p. 41. Available at: https://hdl.handle.net/10568/90133.

Quisumbing, A., Meinzen-Dick, R., Raney, T., Croppenstedt, A., Behrman, J. and Peterman, A. (eds.) (2014). *Gender in agriculture: Closing the knowledge gap*. Dordrecht: Springer Science & Business Media B.V.

Quisumbing, A., Rubin, D., Manfre, C., Waithanji, E., van den Bold, M., Olney, D., Johnson, N. and Meinzen-Dick, R. (2015). Gender, assets, and market-oriented agriculture: Learning from high-value crop and livestock projects in Africa and Asia. *Agriculture and Human Values*, 32(4), pp. 705–725.

Ragot, M., Bonierbale, M. and Weltzien, E. (2018). *From market demand to breeding decisions: A framework*. Lima (Peru). CGIAR Gender and Breeding Initiative. GBI Working Paper. No. 2. Available at: www.rtb.cgiar.org/gender-breeding-initiative.

Rocheleau, D.E. (1995). Gender and biodiversity – A feminist political ecology perspective. *IDS Bulletin-Institute of Development Studies*, 26, pp. 9–16.

Rubin, D. and Manfre, C. (2014). Promoting gender-equitable agricultural value chains: Issues, opportunities, and next steps. In: *Gender in agriculture*. Dordrecht: Springer, pp. 287–313.

Sanya, L., Ssali, R., Akankwasa, R., Nowankunda, K., Barekye, A., Gorreth Namuddu, M. and Kubiriba, J. (2018). Gender-differentiated preferences in breeding for new *matooke* hybrids in Uganda. In: H.A. Tufan, S. Grando and C. Meola, eds., *State of the knowledge for gender in breeding: Case studies for practitioners. Lima (Peru)*. CGIAR gender and breeding initiative. Working Paper. No. 3. Available at: https://cgspace.cgiar.org/handle/10568/92819.

Seguino, S. and Were, M. (2014). *Gendered perspectives on economic growth and development in Sub-Saharan Africa* (No. 2014/056). WIDER Working Paper.

Seymour, G., Doss, C., Marenya, P., Meinzen-Dick, R.S. and Passarelli, S. (2016). *Women's empowerment and the adoption of improved maize varieties: Evidence from Ethiopia, Kenya, and Tanzania*. Paper prepared for presentation at the Agricultural & Applied Economics Association's Annual Meeting, Boston, MA, July 31–Aug. 2, 2016. http://purl.umn.edu/236164.

Smale, M. (2011). *Does household headship affect demand for hybrid maize seed in Kenya? An exploratory analysis based on 2010 survey data. Michigan State University Working Paper 115*. East Lansing, MI: Michigan State University.

Teeken, B., Olaosebikan, O., Haleegoah, J., Oladejo, E., Madu, T., Bello, A., Parkes, E., Egesi, C., Kulakow, P., Kirscht, H. and Tufan, H.A. (2018). Cassava trait preferences of men and women farmers in Nigeria: Implications for breeding. *Economic Botany*, pp. 1–15.

Walker, T., Arega, A., Ndjeunga, J., Labarta, R., Yigezu, Y., Diagne, A., Andrade, R., Andriatsitohaina, R.M., De Groote, H., Mausch, K., Yirga, C., Simtowe, F., Katungi, E., Jogo, W., Jaleta, M. and Pandey S. (2014). *Measuring the effectiveness of crop improvement research in Sub-Saharan Africa from the perspectives of varietal output, adoption, and change:*

20 crops, 30 countries, and 1150 cultivars in farmers' fields. Synthesis Report for Objectives 1 and 2 of Bill & Melinda Gates Foundation's Diffusion and Impact of Improved Varieties in Africa (DIIVA) Project. Rome, Italy: CGIAR Independent Science and Partnership Council (ISPC) Secretariat. Available at: http://impact.cgiar. org/files/pdf/ISPC_DIIVA_synthesis_report_FINAL.pdf.

World Bank. (2012). *World development report 2012: Gender equality and development*. Washington, DC: The World Bank.

Yiadom-Boakye, E., Owusu-Sekyere, E., Kwame Nkegbe, P. and Ohene-Yankyera, K. (2013). Gender, resource use and technical efficiency among rice farmers in Ashanti region, Ghana. *Journal of Agricultural Economics and Development*, 2(3), pp. 102–110.

Yokouchi, T. and Saito, K. (2016). Factors affecting farmers' adoption of NERICA upland rice varieties: The case of a seed producing village in central Benin. *Food Security*, 8(1), pp. 197–209.

3 "Change in the Making"

1970s and 1980s building stones to gender integration in CGIAR agricultural research

Margreet van der Burg

Introduction

> *Development activities cannot achieve their full potentials if women's potentials continue to be neglected. (. . .) Concrete examples (. . .) show that it is possible and not unusually difficult to design and implement development activities that include attention to women's productivity and social roles. Development activities cannot have much value for any of us if they do not have a positive value for women.*
>
> – (Zeidenstein, 1978, p. 971)

Ample 40 years after the 1978 lecture of George Zeidenstein, we still consider his statement valuable and quite contemporary. It makes one wonder what happened in between and what can be done to better strategize and consolidate efforts in the future. This chapter examines this for international agricultural research and development (IARD) by zooming into the CGIAR. Established in 1971 as a worldwide Consultative Group and 2010 as Consortium, CGIAR has become key in IARD with its wide-ranging public international agriculture-related research centers (IARCs) worldwide.

Currently, the CGIAR has a Gender Strategy that was launched in 2011 (CGIAR CB, 2011). In the first gender "Change in the making report" of 2015 the authors recalled CGIAR's commitment to: "*a rapid transition in our research from merely enumerating the women affected by science-based innovations to a more comprehensive and targeted approach for bringing positive change to women as well as men*" (Russell et al., 2015, p. 3) From 2012 onwards, mainstreaming gender into all new CGIAR's Research Programs (CRPs) as a cross-cutting theme, gender capacity-building, and allocating up to 14% of the program budget followed (CGIAR Consortium, 2012; Russell et al., 2015; CGIAR-IEA, 2017; comp. Karlsson and Russell, 2017).

This initiative of 2011 was not CGIAR's first. Though, a 2010 gender scoping study commissioned by the CGIAR concluded in clear language that gender had been integrated poorly so far: "*In spite of a number of strategic gender initiatives, (. . .) most CGIAR Centers historically have not had a clear gender policy, have not mainstreamed gender into the research program or conducted strategic gender research (e.g., gender initiatives), have not trained staff in gender analysis and have not consistently published gender-specific research findings*" (Kauck, Paruzzolo and Schulte, 2010, p. 7). It listed three exceptional examples: (1) the Women in Rice Farming Systems Program (IRRI 1986–1996), (2) Intra-household Program (IFPRI 1992–2003) and (3) Participatory Research and Gender Analysis Program (PRGA), established in 1997 (Kauck, Paruzzolo and Schulte, 2010).

Obviously, the CGIAR did not start at zero in 2011, but there is no up-to-date historical analysis to position the development and integration of gender research into the CGIAR, nor into (former) agricultural faculties and universities, nor in other IARD institutions or IARD as such. Starting with zooming into the CGIAR makes sense since it is associated with all main IARD institutions. Together, the CGIAR institutes cover a wide variety of agri-food systems that besides food crops also includes livestock, forestry, fisheries and aquatic agriculture and natural resources and water management.

This chapter presents the first part of a work in progress based on research on CGIAR gender activities of the 1970s and 1980s. Online searching was considerably easy due to CGIAR's policy of transparency and open access according to its public function. Their sources included publications, working papers, and conference proceedings with participant lists and transcripts of presentations, discussions and recommendations besides assessments, policy documents and meeting reports. Together they give insight into how highly controversial issues were debated, for instance on social equity, ecological impacts as well as gender research and its integration.

Entry points were found through tracking references backward in time, especially in CGIAR gender-related documentation (e.g., IRRI, 1985; Jiggins, 1986; Feldstein and Poats, 1990; Poats, 1990, 1991; CIAT, 1991; Mowbray, 1995; Feldstein, 1995; Feldstein and Merrill-Sands, 1996; Feldstein, 1998, 2000; Doss, 1999, 2001; Paris, 2000; Jackson, 2005; CGIAR, 2008; Kauck, Paruzzolo and Schulte, 2010; PRGA, 2011; Quisumbing et al., 2014). Other documentation was searched on wording such as "women, female, and gender."[1]

This chapter first clarifies the CGIAR mandate and its response to critiques on the so-called *Green Revolution* for its unexpected lower farm yields and increased social inequality. New approaches to "re-people" agriculture in the CGIAR will be explained as niches to integrate women and gender in research. Second, CGIAR gender "milestones" of the 1980s are introduced and analyzed as enabling various gender approaches, and eventually programs up to the 1990s. A CGIAR gender timeline closes this section. Third, the various gender approaches will be distinguished and qualified with a framework based on an adapted version of the Gender Integration Continuum towards transformative gender change (Aregu et al., this volume) and another graph on methodological distinctions. At last, lines for further research and debate on the options and limits of progress toward transformative gender change are elaborated.

Mandate: from green revolution to diversified food and nutrition security

Global food insecurity, maximizing yields and agricultural innovation

For many years, food security and maximization of yields have been coupled within agricultural innovation efforts. Especially in periods of large regional conflicts, natural disasters or exponential population growth, the threat or actual experience of hunger has often urged farmers to increase yields. In the twentieth century, states agreed on intergovernmental cooperation after wide-ranging devastating world wars, first within the League of Nations (1920) and later the United Nations (1945). The extending international community envisioned to achieve a better world for all by securing human rights and global solidarity. The UN formulated food rights as part of human rights and repeatedly also called for social justice and inclusion with regard to vulnerable groups such as women, children, youth and poor, especially those in disadvantaged positions by their race/ethnicity, class, religion or lately health status and sexuality.

Food instability around the two world wars alarmed the international community to better strategize food production worldwide. The establishment of the UN FAO (1945) and later IARCs (1960 onward) marked the beginning of a new era of fighting hunger and malnutrition at a global scale. Critical studies on postwar IARD have also pointed at the geopolitical motivation. By winning hearts against communist expansion, global control over global safety, stability and economies could be established (Anderson, 1991, 2003; Anderson, Levy and Morrison, 1991; Perkins, 1997; Cullather, 2003/2015, 2004, 2010; Smith, 2009; Harwood, 2012). New extrapolations on population growth, shortage of fossil resources and damage on natural resources brought the discussion regularly back onto the international agenda ever since. Science-based agricultural development has been considered best to meet national and international food production potential. It would rather combine outcomes of breeding and agronomic research with macroeconomic checks (e.g., Cépède, 1984; Kortright, 2012).

In 1960, the International Rice Research Institute (IRRI) at the Philippines started with funds of the U.S.-based Rockefeller Foundation (RF) and Ford Foundation (FF) as the first IARC. The IARCs consolidated and strengthened their global public function in 1971 by joining forces in the CGIAR that formalized their cooperation with World bank, FAO and many international and national funding agencies (Chandler, 1982). Increasing yields in food production along the motto "*the more to yield, the more to eat*" became their main focus. Plant breeding was considered crucial to make grain varieties available that would deliver more yield per plant, especially for staple crops such as rice, wheat and maize. New so-called high-yield varieties (HYV) were developed to break the yield ceiling of traditionally bred varieties. To get the most out of the new varieties, these were packaged with new soil and disease management methods supported by irrigation improvement, new equipment and chemical fertilizers, herbicides and pesticides.

IRRI's first success was the HYV rice variety IR-8, launched in 1966, and soon called "*miracle rice*" because of the high yields in test fields. It has been championed as a milestone that encouraged further breeding efforts. (Chandler, 1982) Nevertheless, first IRRI director-general Robert Chandler expressed his distress about an undeniable miscalculation in the so-called Green Revolution: "*On retiring from IRRI in 1972, the only real disappointment I felt was that somehow we did not understand sufficiently why the Asian farmer who had adopted the new varieties was not doing better*" (IRRI, 1979, p. 1). The gap between actual and potential yield of new varieties was later referred to as the yield gap (Datta et al., 1978).

Gender niches: "re-peopling" agriculture in the "Green Revolution"

With questioning why new varieties did not meet the yield expectations, the attention was turned to the people on the fields and farms. This coincided with a growing call for global solidarity with the right to food but also self-determination, especially regarding less privileged groups and future generations. Growing evidence of differential distribution of benefits and not foreseen damaging social and environmental impacts provoked questioning the prevailing technology approaches. CGIAR scientists and administrators found themselves challenged to engage; lively discussions followed to identify and achieve possible improvement. The IARCs started to "re-people" their understanding of agriculture. Claims for advancing social and women's equality got momentum by the global UN Year and later Decade for women (1975–1985).

Impact studies: yield constraints and "re-peopling" agriculture

IRRI started to research and openly discuss yield gaps in the 1970s. After four years of research in seven Asian countries (Datta et al., 1978) the findings were discussed at the 1978 conference *Farm-level Constraints to High Rice Yields in Asia: 1974–1977* (IRRI, 1979). This first CGIAR impact and diagnostic study was led by a multidisciplinary team (economist, agronomist, statistician). It examined the differentiation in adoption of new rice technology, production and yields to reveal the factors causing yield gaps. Biophysical, socioeconomic and institutional constraints were found; the social-economic constraints also included tradition and attitudes, knowledge, input availability and institutions (Datta et al., 1978). It reflects a turn in focus to *who* was innovating or not, and *under what conditions*.

System approaches to support less privileged groups in farming

CROPPING SYSTEMS APPROACH AND USER PERSPECTIVE: TO REACH SMALL FARMERS IN LESS FAVORABLE ENVIRONMENTS

Social consequences of new technologies were already discussed and addressed at IRRI in 1969; it announced the intent to broaden its research approaches when preparing for its second decade (Oasa, 1985). In a policy paper of 1973 on intensifying the small program on cropping systems, IRRI Director-General Nyle Brady with others stressed to better align innovation to "*the cultivators' world*" to tackle environmental, institutional, and socioeconomic constraints (Brady et al., 1973). They stated that IRRI had to change itself instead of pointing at social resistance as a limiting factor as once assumed. This way IRRI broadened its focus on yield of staple food crops under optimal circumstances to smallholding mixed-farming practices in less favorable circumstances. Although farmers were not part of a crop system, it did enable researchers to differentiate in target groups, thus also women, and introduced a so-called *user perspective* into research considerations.

INTEGRATED RURAL DEVELOPMENT: TO INTEGRATE SMALLHOLDERS AND THE RURAL POOR IN DEVELOPMENT

Oasa stressed that CGIAR took part in a more general trend of "systems" approaches. He pointed at director Montague Yudelman of the World Bank, to which the CGIAR was aligned since 1971. Yudelman in 1976 revalued the small-farm sector by no longer framing it as merely providing labor surplus for industrialization but taking it seriously in its existence as producer of agricultural surplus. As result, the World Bank broadened its small-farm program with agriculture-oriented technological innovation. But to integrate the rural poor into rural development, his program went beyond agriculture by providing ample opportunity to education, health and nutrition (Oasa, 1985, ref. to Yudelman, 1976). This perspective enabled the program to include all rural inhabitants, thus also women.

FARMING SYSTEMS APPROACH: HOLISTIC AND MULTIDISCIPLINARY

Further widening the scope in the mid-1970s to a whole-farm orientation led to the Farming Systems Approach (FSA). In 1978, this new system approach was discussed at a CGIAR-wide conference to define its possible integration into CGIAR research. FSA was especially profiled by its holistic and multidisciplinary character. Basically, FSA considered farming as an open system that was organically tied to many internal and surrounding systems. FSA prompted a problem-oriented approach that systematically traced occurring issues or constraints back to wider farm constellations and therefore often found unexpected locations for possible solutions (CGIAR TAC, 1978). Its approach intrinsically acknowledged that change in one part definitely would affect other parts in multiple and often unforeseen ways. Consequently, examination of possible (later called ex-ante) impacts besides actual (ex-post) impacts was suggested. David Norman explicitly urged to assess all relevant factors and aspects before recommending any technological improvement. He suggested a multidisciplinary perspective that would include a "*human dimension*" and distinguish endogenous aspects such as land, labor, capital and management, beside exogenous ones as community structure, norms, beliefs, extension and credit and inputs distribution systems (Norman, 1978). This resonated with the 1978 review that FSA would "*improve our understanding of the farmer, his skills, preferences and aspirations*" (CGIAR TAC, 1978, p. 21). It made way for participatory methods and including also women as part of farming constellations. More details on FSA and its constraints to integration can be found in Edmund Oasa (1985) and Mike Collinson (2000).

Multi- and interdisciplinarity: the inclusion of social scientists and new participatory methods

The recognition of farming people as human factors and thus actors of agricultural change led the CGIAR to include more disciplines of social scientific research than economics. Norman with others expressed social scientists would best take up the anticipating constructive role (Oasa, 1985 ref. to Norman et al., 1982). Already from 1974 onward, recently graduated social sciences PhDs were contracted for two years under a Rockefeller Foundation fellowship program (1974–2001) (Conway et al., 2002). These so-called *Rocky Docs* were: "*to prepare future generations of first-rate scholars to staff the social sciences component of the growing international agricultural research network*" (Moock, 1989, p. ix). Many young anthropologists and some women or gender studies scientists joined and especially elaborated new on-farm participatory methods (e.g., Collinson, 2000; Gibbon, 2012). The Rockefeller Foundation assessed it as a success after 15 years (Moock, 1989, p. ix). A 1988 CGIAR-wide assessment affirmed that socioeconomic research had been undertaken in most CGIAR centers, especially within farming systems research (Anderson, Herdt and Scobie, 1988). Remarkable though was the warning that openings in social scientist positions were required in order to gain continuous effect (Moock, 1989, p. ix) while both gender and other social scientists kept facing difficulties around their integration.[2]

CGIAR gender milestones: inclusion of women and gender

Mentioning of women without analysis or policy, 1970s/early 1980s

1970s: UN decade of women and CGIAR mentioning women without analysis or policy

The well-known book of Ester Boserup (1970) and the UN Decade for Women (1975–1985) spurred the attention for integrating women and gender in research rapidly. FAO agreed in 1975 to fully integrate women in its rural development programs and projects. Except of building on the division of Home Economics and Social Programmes Service FAO announced to promote mainstreaming this into all programs, divisions and services. It illustrates the transition in the 1970s from addressing women by (agricultural and rural) home economics unto feminist, women's and gender studies as a change to Women in Development (WID-1970s) and later Gender in Development (GAD-1980s). These did not wipe out but merely overshadowed the earlier welfare approach (Moser, 1989). This shift changed the focus from research "about and for women" toward self-determination "by women and on their own terms" (WID), beyond traditional women's spheres and complementary gender roles as often defined in existing Western family ideologies.[3]

The need and practice of including women and gender in IARD were gradually supported by many governments, universities and foundations from the U.S. (1973 Percy amendment) and other countries (e.g., Poats, 1990; van der Burg and Bos-Boers, 2003; Fraser and Tinker, 2004; Kauck, Paruzzolo and Schulte, 2010). Though, despite its connections, the CGIAR documentation of the 1970s only accidentally mentioned women without any analysis. A first example dates from discussions on social-economic aspects of IARD in 1973. From the FAO contribution, it was noted that the roles of women and children had to be included (TAC meeting, 1973). Other examples are a statement that the family was a key factor for successfully introducing new farming systems (CIAT, 1973) and an observation of having many female-headed households in the research area (Bradfield, 1974). A last example concerns IRRI that in 1976 under leadership of later DG but then TAC chair M.S. Swaminathan was reviewed. IRRI was urged to help upgrade the technical skills of women workers and expand training programs for women, both in adaptive research and field extension (CGIAR TAC 1976, pp. 82 and 93). Nevertheless, these observations did not lead to specific policies or research.

Early 1980s: no CGIAR policies but urgency expressed in reviews and item for discussion

In 1980/81 women entered the CGIAR system-wide agenda for the first time. In line with growing impact concerns, the CGIAR decided to also assess "*the value of the system to the most disadvantaged groups, including resource-poor farmers and women*" (Consultative Group Meeting Oct 1980, p. 22). Indeed, the 1981 Second Review of the CGIAR underlined under Human Resources "*the role of women in the system*" that women played an important role in agricultural production as farm owners, managers, sales agents and field workers. It stated that overlooking women had led to reduced impact or even total failure of programs. Therefore, the IARCs should ensure including the role of women in the technology-generation process and analyzing possible differential benefits for men and women. In particular, farming systems programs were considered the more suitable. But the recommendations stressed to have more women in trainings and, as owing to the important role of women in agriculture in many developing countries, to increase the participation of women as professional staff and in CGIAR bodies. These were supported as consistent with most major agencies (CGIAR Second Review, 1981, pp. viii, xvii, 36, 97; Committee's Report, November 1981, added to review, p. 3).

In 1983, Philippine sociologist Gelia Castillo was appointed as first woman chair of the Board of Trustees of CIP. But when the Women in Development Network of the Midwestern Universities Consortium for International Activities (MUCIA) allowed to present a proposal to develop consultation and educational services on women's productive roles in agriculture for the CGIAR in the same year, it withheld its support (Poats, 1990, 1991; Paris, 2000).

Finally, in 1984, Janice Jiggins was commissioned to assess the gender-related impacts for the 1982–1988 CGIAR-wide assessment (Jiggins, 1986). Assessing the benefits for women in farm families was included to address the criticism on the Green Revolution that it had been "*prejudicial to the interests of the poor*" (Farrar in Anderson, Herdt and Scobie, 1988, foreword). Jiggins concluded that disregarding gender issues in agricultural research and technology development would hold output, productivity and welfare below the potential. If women's roles in production and food systems were not taken seriously, the CGIAR goals could not be reached, she argued (Jiggins, 1986). Despite her rich analysis and exposure of many IARD aspects of importance to women, only one box, *Women and Farming Systems Research*, appeared as stand-alone in the final report (Anderson, Herdt and Scobie, 1988, p. 48). The impact chapter added examples from gender research but without any references. It exemplified aspects of class, ethnicity and gender in productivity, allocation of land and tasks and intra-household gain and distribution of income and food. Almost as an excuse, it was stated that centers were just slow with addressing women (Anderson, Herdt and Scobie, 1988, ch 8) Nevertheless, they were said to be progressing in raising awareness, developing technological improvements and avoiding innovations that might harm women's welfare (Anderson, Herdt and Scobie, 1988).

Women/gender in farming systems and user perspective, 1980s

1983 gender milestone: IRRI conference Women in Rice Farming Systems

The *Women in Rice Farming Systems Conference*, organized by IRRI in 1983, was this first CGIAR women's or gender conference with 78 participants from 27 countries. The book *Women in Rice Farming* is the tangible result of new international connections. This real treasure full of rich articles on women in African and Asian rice-based farming addressed agricultural activities, non-farming income-providing activities, and family, community and household care activities. Recommendations agreed upon by the participants were included as an annex (IRRI, 1985).

Taking farming systems as the conference focal point reflected its engagement with this new people-based approach. Many presenters included impact research but mainly newly focused on social impacts, for instance on women's labor, related to the very complex contexts of their lives. Conference participants also emphasized the importance to differentiate among women by including other social dimensions than gender, especially class, status in family, religion, and contexts defined by (out) migration and changing labor markets (IRRI, 1985). The complexity of African rice farming systems was very well illustrated by Jennie Dey. She highlighted how social norms besides physical and biological aspects affected African farm systems and practices while extending production with consumption aspects and addressing both the household and village levels. Based on irrigation and (re)settlement projects, she evidenced that not considering family structures would readily leave women with less control and opportunity to provide as used and expected to, thus weakening their self-determination and gender status (Dey, 1985).

CONFERENCE GENDER RESEARCH RECOMMENDATIONS

Besides an explicit recognition of the pivotal role of women in advancing agricultural progress and agrarian prosperity, a clear call in the conference recommendations was made to actively involve women in all innovation stages of farming systems programs while explicitly harnessing their success and avoiding unanticipated hardships. Another recommendation questioned maximizing rice yields as IRRI's general goal of technology development, as it could easily conflict with income effects. The participants also propagated ex-ante impact studies to assess technology impact on the employment of women in advance, regarding both wage rates and women's agricultural tasks. Priority was given to farm and domestic labor-saving technologies that would improve real wages and income (IRRI, 1985). Noticeable is that not only women in agriculture but all rural women were seen as impacted thus to be included. At last, the conference participants advocated for widening the current methodology and scope of scales in constraints research. They urged to also consider social, educational, and institutional constraints and to add the household level as crucial for not missing key differences (IRRI, 1985). Thus, both the old and relatively new people-based approaches were to be changed in their view.

WOMEN IN RICE FARMING SYSTEMS PROGRAM (WIRFS) UNDER THE LEAD OF IRRI

The conference resulted in establishing the Women in Rice Farming Systems Program under leadership of IRRI. M.S. Swaminathan presented its three pillars with at first impact research on women, especially in small family farming and farm wage labor. The second pillar was to design, test, and adapt technologies to reduce "drudgery" as well as increase women's productivity and income. This included diversifying opportunities for gainful employment both on-farm and off-farm. Drudgery was associated with inferior work quality when work was repetitive, concerned routine only and required practically no skill. The third was addressing gaps in skill, input delivery systems, government policies and whatever else hampered the full participation of women in technology development and adoption (Swaminathan, 1985). His words framed the program into the overall mission of IRRI while addressing women as a special target group. Though he did not underestimate the required change: "*The greatest challenge (. . .) lies in motivating scientists and technologists to (. . .) 'listening and learning' through collaboration with poor women while developing their research priorities and strategies*" (Swaminathan, 1985, p. 60).

The groundwork for the program was done by Jenny Dey, who was temporarily hired from FAO. She saw to enrich the farming systems approach by redressing the household as a homogeneous unit, looking beyond output and productivity as sole innovation effects and advancing interdisciplinarity in research programs (Dey, 1985, 27 Aug., IRRI archives, box 40). Finally, program principles were agreed upon with national agricultural researchers of Bangladesh, India, Indonesia, Nepal, the Philippines and Thailand in line with Swaminathan's wording: to actively involve biophysical scientists, involve women in all stages of technology development and transfer and generate on-farm and off-farm employment for women (Dey, 1985, 27 Aug., IRRI archives, box 40). The WIRFS program engaged with participatory methods to secure an inclusive approach to women and was included in the Asian Rice Farming Systems Network (ARFSN) to secure interdisciplinarity.[4]

New research methodologies and capacity building were taken up in 1988. Being well connected to gender researchers abroad, Hilary Feldstein and Cornelia Flora were invited to train in gender analysis. They used an IRRI case study (Paris, 2000) that was later included in *Working Together. Gender Analysis in Agriculture* (Feldstein and Poats, 1989). Technology development was addressed in a 10-day travelling WIRFS seminar through China in 1989. On-site demonstration introduced "*simple*" technologies to ease women's farm, domestic or processing labor (Zandstra, 1989). Later, the 1994 seminar discussed when and how technologies would help or hurt rural women. Thelma Paris and Prabhu Pingali set the stage, after which many interesting nuanced contributions followed (Quick and Paris, 1995).

The first CGIAR review of the WIRFS program occurred within the IRRI review of 1987 and was very supportive. IRRI consequently committed to advance the work within its own and collaborative research projects (Paris, 2000). After five years of WIRFS in 1990, IRRI asked gender scientists Susan Poats and Hilary Feldstein for an expert review. They recommended to move to a "gender" perspective and focus more on gender constraints to productivity and technology development. This would better allow them to involve more scientists and integrate gender analysis and impact assessment into more specialized areas of strategic research, such as seed management, varietal improvement, integrated pest management (IPM), and biomass utilization (Paris, 2000, ref. to Feldstein and Poats, 1990). Although all points were taken up, in 1995, the WIRFS stopped. The Ford Foundation funding expired, and no one had succeeded in securing management and financial support (Paris, 2000).

Michael Collinson and Hilary Feldstein in 1995 stated that WIRFS had moved "*beyond counting of hectares of increased yield.*" They emphasized that skeptical scientists at IRRI seemed to have been turned around. First seen as work in service of the national partners, they had admitted gender analysis had also benefitted their work (Mowbray, 1995, p. 3). More details on WIRFS can be found in Chater and Carangal (1996, pp. 41–46).

1985 Gender Milestone: intercenter seminar women and agricultural technology as "users"

After the 1983 IRRI conference, the first CGIAR-wide gender event followed in 1985: the intercenter seminar *Women and Agricultural Technology: Relevance for Research* at Bellagio, Italy. In CGIAR circles, it was presented as starting a system-wide dialogue on women and agricultural development (CGIAR News, 1985). An objective was set to assess current activities and seek more effective integration of women in the modernization of agriculture and a better CGIAR performance in this. Among the participants were upper-level CGIAR managers and stakeholders. Ester Boserup led one of the panels. M.S. Swaminathan prepared recommendations to discuss at the end. The full report included a concluding statement (RF and ISNAR, 1985).

This was the first time CGIAR assessed, connected and presented research on women as a concerted CGIAR effort. The final statement called for a long-term strategy that would involve the "*consideration of women, where possible, in all phases of its research and development works, including feedback from female farmers*" (RF and ISNAR, 1985, Vol 1, p. 2). It also affirmed gender as "*an important variable in distinguishing among potential beneficiary groups for agricultural technology research and policy analysis*" (RF and ISNAR, 1985, Vol 1, p. 1; underlined by author). The wording surely mirrored the angle of technology developers, as did the continuously adding "*other beneficiaries of research*" or "*other users of technology*" when addressing women in agriculture. Also were women's needs and potential for change highlighted as restricted by their specific women's "*unique responsibilities in child care and household maintenance*" (RF and ISNAR, 1985, Vol 1, p. 2). The formal concluding statement thus took the user perspective on women as a specific target group among other users. Nevertheless, the seminar also had contributions that connected women's constraints to their position in agriculture, agricultural innovation processes and technology adoption.

Programmatic approach from the 1990s onwards

Late 1980s: discussions and assessments 1987–1990

With the 1985 statements in hand, the CGIAR reviews indeed started to assess the IARCs' gender efforts. For instance, in the review meetings of the International Centers Week (ICW) of 1987, IFPRI was urged to do more on women (CGIAR, 1987, p. 7), while IRRI was encouraged to retain the work on women in rice farming (ibidem, pp. 17–18). The same ICW included a seminar on gender issues: *User Impact, Agricultural Technology and the Global Agricultural Research System* (CGIAR, 1987, Appendix I). The reported outcomes were consistent with the 1985 intercenter seminar and once again stressed to monitor progress on a CGIAR-wide routine basis and assess gender integration in external reviews of IARCs. The usefulness of a standard-setting review was raised. When this was repeated by donors in 1989, Susan Poats was asked to write and present the second CGIAR gender assessment in May 1990 (Poats, 1990, published in 1991).

These first CGIAR efforts and deliberations ultimately resulted in three CGIAR programmatic initiatives that put gender research on the CGIAR agenda on a system level. Further investigation of this new stage unfortunately has to wait. The programs were:

1991 GENDER MILESTONE: PROGRAM GENDER ANALYSIS, 1991–1995

(see esp. Poats, 1990, 1991; CIAT, 1991; Feldstein, 1995, 1998, 2000; Feldstein and Merrill-Sands, 1996)

1992/1996 GENDER MILESTONE: PROGRAM INTRA HOUSEHOLD ANALYSIS, 1992/1996–2005

(see esp. Jackson, 2005)

1996 GENDER MILESTONE: PROGRAM PARTICIPATORY RESEARCH AND GENDER ANALYSIS (PRGA), 1996–2011

(see esp. CGIAR-PRGA, 1997; PRGA, 2011)

Timeline of CGIAR gender milestones (see Figure 3.1)

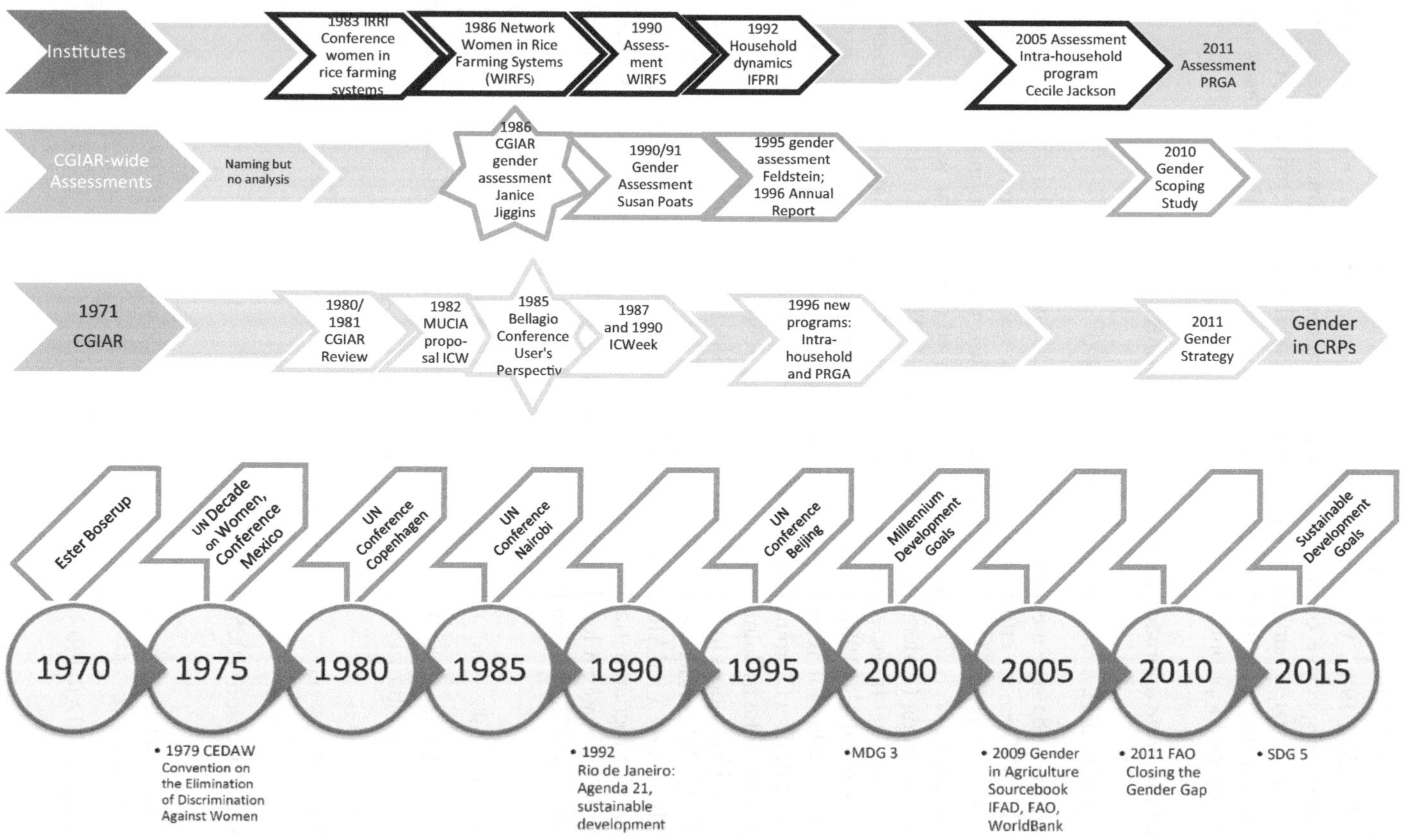

Figure 3.1 Timeline of CGIAR gender milestones

Further research on niches and obstacles in CGIAR mandate and practice

Qualifying various gender approaches within a gender integration continuum

Obviously lower uptake than expected and social equity issues around the "Green Revolution" led to include new approaches in IARCs in the 1970s. These pushed agriculture as a socially embedded sector and farming as social performance, and were therefore key to the "re-peopling" of agriculture and inclusion of social science aspects. The new impact and diagnostic studies, integrated rural development and farming system research opened options to women and gender integration. From the 1980s onward, specific women and gender aspects and approaches were articulated, debated and contested. The following identifies what specifics are to be further uncovered and addressed.

In line with current insights of gender studies in agriculture Figure 3.2 further extends the gender integration continuum as adapted in Aregu et al. (this volume). First, it differentiated between gender-blind and gender-aware approaches. Of the gender-aware approaches the gender accommodating approach in principle takes the existing gender situation as a start or entry point, for instance to avoid conflict and smoothen collaboration. In Aregu et al., we illustrated that such an approach easily reinforces unequal social and gender relations or even worsens them. Where it effects in more women's agency and access to resources it can though become a step forward toward gender-transformative change. Though this may not be sustainable if wider-ranging change and consolidation are not anticipated. A gender-transformative approach can take the same entry point though it explicitly aims at stretching towards advancing transformative gender change (Aregu et al., this volume).

My suggestion is to develop qualifying markers that are crucial for in-depth gender analysis in support of gender-transformative research in IARD. Therefore Figure 3.2 lists a combination of methodologies most associated with transformative gender IARD approaches, stretched to the most in-depth level, in line with recent literature (e.g., Currie and Vernooy, 2010; Quisumbing et al., 2014; Hillenbrand et al., 2015; IGWG, 2017; van Eerdewijk et al., 2017; Petesch et al., 2018). Braithwaite (2001, p. 5) suggested the six levels of analysis and Cho et al. (2013) address intersectionality.

Interesting is that in the early days of CGIAR women and gender integration, most all levels of analysis were mentioned and debated. We noted that concepts as labor, production, productivity, household, family, labor market and economic growth were questioned and investigated. Heterogeneity and interlocking of systems were addressed and discussed (e.g., CGIAR related: Dauber and Cain, 1981; IRRI, 1985; Moock, 1986; Poats et al., 1988). Often Western male bias was tackled. With a sharp eye for cultural aspects especially, social scientists trained in ethnographic methods challenged many assumptions with empirical accurateness; they connected manifold women's positions to complex social organizations

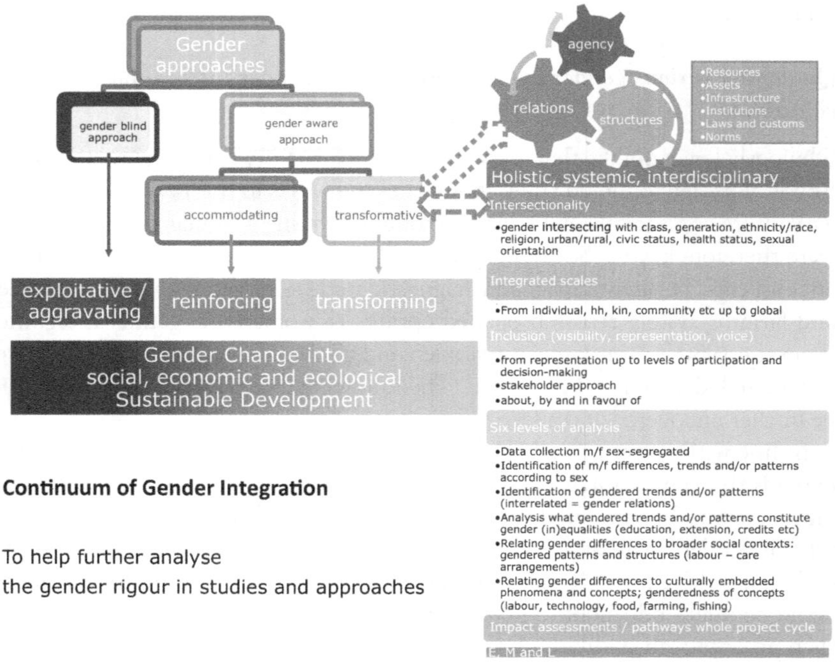

Figure 3.2 Gender integration continuum with qualifiers for transformative gender research

around farming and the gender differences in material circumstances. Also, the methodological books with case studies are impressive results of CGIAR's gender efforts to develop and train gender-sensitive methodologies, concepts and analysis (e.g., Overholt et al., 1985; Feldstein and Poats, 1989; Feldstein and Jiggins, 1994). These are remarkably interesting for contemporary use still!

Gender integration and distinction of component and systemic approaches

Although most of the list in Figure 3.3 was addressed in early CGIAR gender research, Cheryl Doss (1999 and 2001) observed that a lot of naivety within the CGIAR had to be redressed as if just including women in research as usual would do the job. Also from the 2010 scoping study (Kauck, Paruzzolo and Schulte, 2010) and the historical analysis so far, we can conclude that there has not been one linear development to progressive integration of gender-systemic or -transformative approaches or change.

Two distinct angles to gender integration within agriculture come to the fore when comparing the 1980s research practices to ongoing issues up to the 2010s. They basically represent either a component or systemic approach. This is in line with the distinction of practical and strategic needs by Moser (1989). As seen, the new 1970s approaches opened opportunities for both. In the 1980s, these are reflected in gender-integration approaches and debates from either a "*user perspective*" (e.g., 1985 intercenter seminar) or system perspective (e.g., FSA as IRRI's WIRFS). Systemic gender integration and gender change in a systemic transformative way though was pushed back by the CGIAR as institutional system through allowing and stressing gender integration as adding a variable or target group, thus as a component. It took the 2010s gender initiatives to get systemic research fully on the CGIAR agenda again.[5]

Figure 3.3 clarifies the distinction between these gender (integration) approaches, although these show extremes that in reality will appear to be more fluid and overlap. The aim is to enable researchers to better locate, identify and examine what favors which gender approaches in what IARD or CGIAR environments. Future research can examine what gender research took similar and different ways to navigate through the institutional context of opportunities and obstacles.

Gender integration and required transformative institutional change

Assessments and scoping studies give pointers for further investigation connected to institutional cultures and practices. From those and through participation within CGIAR activities, it can be concluded that gender research is often

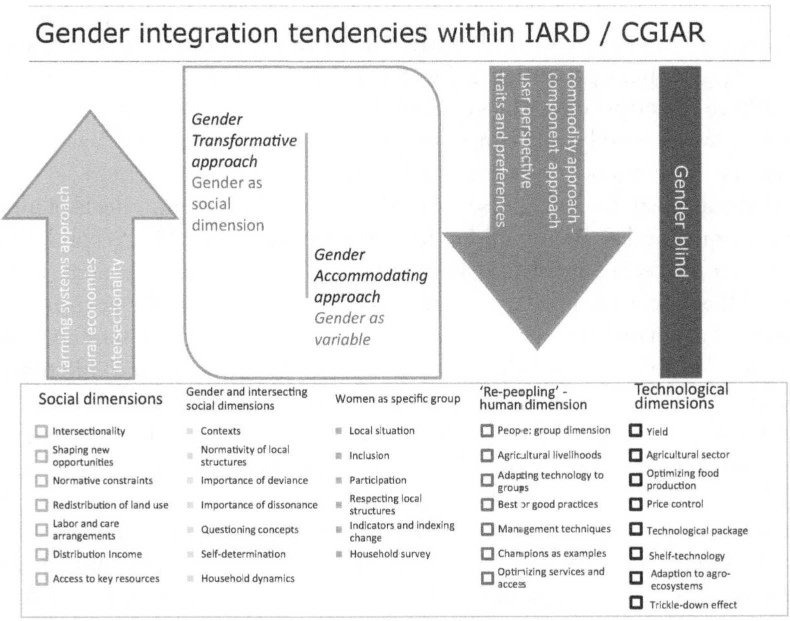

Figure 3.3 Gender integration positions within IARD/CGIAR

looked upon as lowest on the ladder of sciences. Recommendations have tried to reverse this hierarchal backlash with high-level support and budget guarantees. In such circumstances, it can also become a strategy to take a research approach more near to the mainstream as best for integration and continuity. Systemic or structural obstacles within the institution thus can frustrate and even sabotage full development of systemic approaches as the transformative gender research approach in many ways.

Early diverging positions toward equity issues and gender change seem still present and to be discussed. Even though participatory methodologies have brought researchers more in contact with practitioners and stakeholders, the selection and weighting of contacts, statements and priorities are to be further discussed (e.g., CGIAR-PRGA, 1997; Gibbon, 2012). Clearly, efforts have to be aligned and momentum of occurring niches have to be taken, strengthened and consolidated in close tuning of institutions, research approaches and development practices (e.g., also Klerkx et al., 2012). This suggests to consider historical and future CGIAR and IARD gender research approaches not as an institutional continuum or sequence but as a wide-ranging spectrum of partly separate as well as partly connected and overlapping. Future IARD gender research development would then better discuss institutional options in relation to what gender approaches can be taken up, parallel, in sequences or apart.

Conclusions

In tracing the first part of the history of gender integration into the CGIAR IARD, we found within a changing CGIAR mandate and responses to the unforeseen and critiqued effects of the Green Revolution that new impact and system approaches opened up niches for gender research. The "re-peopling" of agriculture in perspectives on agricultural innovation allowed researchers to also integrate women and gender in research on impacts, user perspectives, farming systems, intra-household dynamics and participatory development.

First, impact studies enabled scholars to distinguish diverging effects of innovations on men and women, and led to using gender as variable. Second, the so-called *user perspective* enabled to develop technology for women as users, as target group, thus also took gender as variable. Thirdly, conceptually acknowledging farmers as agents within their livelihood strategies through the farming systems approach led to also addressing women and gender in broad and differentiated farming constellations, eventually combined with a user perspective. Especially within this stream, gender research questioned unitary economic approaches of the (farm) household and fostered agriculture specific gender analysis and accompanying methodologies. The rich literature from this early period is still very valuable. CGIAR gender research developed within an encouraging international context. It shows in-depth empirical research for understanding manifold gendered aspects and impacts as well as shaping new connections, angles and methodologies beyond many earlier assumptions and unseen aspects.

Despite the great early gender milestones, CGIAR had to take up a system-wide gender strategy in 2011 beyond enumerating women. It amazingly reflected many of the 1980s recommendations. To support understanding and working on gender IARD, two graphs were introduced to help to distinguish and qualify different gender approaches. Gender accommodating approaches basically aim at gender inclusion, focus on one component such as preferences in crop varieties and often just take gender as a variable. They risk reinforcing gender inequalities. Gender-transformative approaches take a system approach by also addressing engrained gendered practices and structures, thus also questioning gender norms up to concepts used such as economic empowerment, productivity and labor. These risk encountering much resistance and consequently to be frustrated, marginalized or even stopped. Further research and debate on the historical and contemporary trends, contexts, options and limits of both types of gender IARD are needed.

Notes

1 Another forthcoming article connects CGIAR gender integration to gender in development literature.
2 Further investigation on how opportunities and constraints for social scientists in general and gender scientists in particular paralleled and differed is recommended!
3 More research on this transition in international farm and rural women's development programming is needed (e.g., comp. Mead, 1976; Flora, 1982; van der Burg, 2002, 2010).
4 See Ph.D. thesis of Thelma Paris for early gender developments at IRRI and CGIAR (Paris, 2000).
5 Diana Lopez and Margreet van der Burg currently write an article defining and addressing risks of "instrumentalization" in IARD/CGIAR; further comparison with social sciences in general is recommended as recommended in note 2

References

Anderson, J.R., Herdt, R.W. and Scobie, G.M. (1988). *Science and food the CGIAR and its partners*. Washington, DC: Published for the CGIAR, The International Bank for Reconstruction and Development/The World Bank.

Anderson, R.S. (2003). *American foundations, the green revolution, and the CGIAR: Intentions, implementation, and contingencies*. Unpublished notes prepared for a conference at the University of Manchester on the Foundations of Globalization, November 2003.

Anderson, R.S. (1991). The origins of the international rice research institute. *Minerva*, 29(1), pp. 61–89.

Anderson, R.S., Levy, E. and Morrison, B.M. (1991). *Rice science and development politics: Research strategies and IRRI's technologies confront Asian diversity (1950–1980)*. Oxford: Clarendon Press.

Aregu, L., Choudhury, A., Rajaratnam, S., van der Burg, M. and McDougall, C. (this volume). Implications of agricultural innovation on gender norms: Gender approaches in aquatic agriculture in Bangladesh. In: this volume, pages 162–179.

Boserup, E. (1970). *Women's role in economic development*. New York: St. Martin's Press.

Bradfield, S. (1974). *Report on a field visit to Cacaotal, January 22–30, 1974*. Centro Internacional de Agricultura Tropical (CIAT), Cali, Columbia. Available at: http://hdl.handle.net/10568/7155.

Brady, N., Athwal, D.S. and Hill, F.F. (1973). *A proposal for broadening the mission of the international rice research institute*. Unpublished paper of 1 July 1973 in IRRI library.

Braithwaite, M. (2001). *Gender in research: Gender impact assessment of the FP5 specific programmes of the fifth framework programme – Improving human research potential and the socio-economic knowledge base*. Tacitus sprl.; EU. European Commission. DG for Research, Luxembourg: EUR-OP 2001; Available at: http://collections.internetmemory.org/haeu/20160318155422/http://cordis.europa.eu/pub/science-society/docs/genfp5_imp_hum_res_pot_sum.pdf.

Burg, M. van der. (2010). Rural women's voices and the identification of rural women's issues within the landscape of international organisations, 1889–1940. In: W. Ernst, ed., *Grenzregime: Geschlechterkonstellationen zwichen Kulturen und Räumen der Globaliserung*. Berlin: LIT Verlag, pp. 204–226.

Burg, M. van der. (2002). *'Geen tweede boer.' Gender, landbouwmodernisering en onderwijs aan plattelandsvrouwen in Nederland, 1863–1968 ('No second farmer'. gender, agricultural modernisation and education for rural women in the Netherlands, 1863–1968)*. Hilversum: Verloren.

Burg, M. van der and Bos-Boers, M. (2003). *Vrouwen, Wageningen en de Wereld. Wetenschap, studie en loopbaan, 1918–2003 (Women, Wageningen and the World. Science, study and career at Wageningen agricultural college/ university 1918–2003)*. Hilversum: Verloren.

Cépède, M. (1984). The fight against hunger. Its history on the international agenda. *Food Policy*, 9(4), pp. 282–290.

CGIAR. (2008). *Bringing together the best of science and the best of development*. Independent Review of the CGIAR System Technical Report.

CGIAR. (1987) Main Conclusions reached and decisions taken, Consultative Group Meeting October 26-30, 1987 Washington, DC. Available at: https://hdl.handle.net/10947/274 [Accessed September 2017].

CGIAR. (1981). *Second review of the CGIAR. Report of the review committee*. Available at: https://cgspace.cgiar.org/handle/10947/1283 [Accessed September 2017].

CGIAR Consortium. (2012). *Meeting report progress on gender and agriculture research in CGIAR*, June 15, 2012, at CIRAD Paris Headquarters Paris, France.

The CGIAR Consortium Board. (2011). *Consortium level gender strategy*. Montpellier, France: CGIAR consortium.

CGIAR-IEA. (2017). *Evaluation of gender in CGIAR. – Volume I: Evaluation of gender in research*. Rome, Italy: Independent Evaluation Arrangement (IEA) of CGIAR.

CGIAR-PRGA. (1997). *New frontiers in participatory research and gender analysis*. Proceedings of the International Seminar on Participatory Research and Gender Analysis for Technology Development, September 9–14, 1996. CGIAR Systemwide Program on Participatory Research and Gender Analysis for Technology Development and Institutional Innovation. CIAT publication 294. Cali, Colombia: CIAT.

CGIAR Technical Advisory Committee – TAC. (1978). *Farming systems research at the international agricultural research centers, I – Analysis by the TAC review team of farming systems; Research at CIAT, IITA, ICRISAT and IRRI, 1–164, and II – Proceedings of the Workshop on Farming Systems Research, Nairobi, May 29–31, 1978*, pp. 165–197 (1–30 original numbering).

CGIAR Technical Advisory Committee – TAC. (1976) *Report on the TAC quinquennial review mission to the international rice research institute*. Rome: TAC Secretariat. Available at: https://hdl.handle.net/10947/1394 [Accessed September 2017].

Chandler, R.F. (1982). *An adventure in applied science: A history of the international rice research institute*. Los Baňos, Philippines: IRRI.

Chater, S. and Carangal, V. (1996). *On farmers' fields: Portrait of a network*. Los Baňos, Philippines: IRRI.

Cho, S., Crenshaw, K.W. and McCall, L. (2013). Toward a field of intersectionality studies: Theory, applications, and praxis. *Signs: Journal of Women in Culture and Society*, 38(4), pp. 785–810.

CIAT. (1991). *Gender analysis in agricultural research*. Proceedings of an internal workshop, 13–14 June 1991, Cali, Colombia, CIAT.

CIAT. (1973). *Ceremony of inauguration of CIAT's new facilities, October 12, 1973 and Symposium on the potential of the Lowland Tropics*, 1973, Cali, Colombia. Available at: http://hdl.handle.net/10568/56341.

Collinson, M. (ed.) (2000). *A history of farming systems research*. Wallingford: CABI.

Conway, G., Adesina, A., Lynam, J. and Moock, J. (2002). The Rockefeller foundation and social research in agriculture. In: M. Cernea and A. Kassam, eds., *Researching the culture in agriculture: Social research for international development*. Wallingford: CABI, pp. 373–381.

Cullather, N. (2010). *The hungry world. America's cold war battle against poverty in Asia*. Cambridge and London: Harvard University Press.

Cullather, N. (2004). Miracles of modernization: The green revolution and the apotheosis of technology. *Diplomatic History*, 28(2), pp. 227–254.

Cullather, N. (2003/2015). Parable of seeds: The green revolution in the modernizing imagination. In: M. Frey, R.W. Pruessen and T.T. Yong, eds., *The transformation of Southeast Asia: International perspectives on decolonization*. 2003 first published by M.E. Sharpe. 2015 published by London and New York: Routledge, pp. 257–267.

Currie, D.H. and Vernooy, R. (2010). Social and gender analysis in development practice: Lessons from the field. *Gender, Technology and Development*, 14(1), pp. 1–23.

Datta, S.K. de, Gomez, K.A., Herdt, R.W. and Barker, R. (1978). *A handbook on the methodology for an integrated experiment – Survey on rice yield constraints*. Los Baños, Philippines: IRRI. Available at: http://books.irri.org/9711040360_content.pdf.

Dauber, R. and Cain, M. (eds.) (1981). *Women and technological change in developing countries*. Boulder, CO: Westview Press.

Dey, J. (1985). 23 Women in African rice farming systems. In: IRRI, *Women in rice farming. Proceedings of a Conference on women in rice farming systems, IRRI, Manila, Philippines, 26–30 September* 1983. IRRI and Gower, pp. 419–445.

Doss, C.R. (2001). Designing agricultural technology for African women farmers: Lessons from 25 years of experience. *World Development*, 29/12, pp. 2075–2092.

Doss, C.R. (1999). *Twenty-five years of research on women farmers in Africa: Lessons and implications for agricultural research institutions; With an annotated bibliography*. CIMMYT Economics Program Paper 102.

Eerdewijk, A.H.J.M. van, Wong, F., Vaast, C., Newton, J., Tyszler, M. and Pennington, A. (2017). *White paper: A conceptual model on women and girls' empowerment*. Amsterdam: Royal Tropical Institute (KIT).

Feldstein, H.S. (2000). Gender analysis: Making women visible and improving social analysis. In: M. Collinson, ed., *A history of farm systems research*. Wallingford, UK:CABI and Rome, Italy: FAO, pp. 67–76.

Feldstein, H.S. (1998). *An inventory of gender-related research and training in the consultative group on international agricultural research (CGIAR) Centers 1996–1998*. Cali, Colombia: CGIAR.

Feldstein, H.S. (1995). *Inventory of gender-related research and training in the international agricultural research centers 1990–1995*. Prepared by Hilary Sims Feldstein with Alison Slack. CGIAR Gender Program Working Paper 8, CGIAR. Washington, DC: Worldbank.

Feldstein, H.S. and Jiggins, J. (1994). *Tools for the field: Methodologies handbook on gender analysis in agriculture*. West Harford, CT: Kumarian Press.

Feldstein, H.S. and Merrill-Sands, D. (1996). Gender and the CGIAR: Inquiry, progress, and challenge. In: *CGIAR annual report 1995–1996. Part one: The year in review*. CGIAR, pp. 41–58. Available at: http://hdl.handle.net/10947/5696.

Feldstein, H. and Poats, S. (1990). *Women in rice farming systems: Review report*. Los Baños, Philippines: IRRI.

Feldstein, H.S. and Poats, S.V. (eds.) (1989). *Working together. Gender analysis in agriculture*. Volume 1 Case Studies; Volume II Teaching Notes. West Hartford, CT: Kumurian Press.

Flora, C.B. (1982). Farming systems research and the land grant system: Transferring assumptions overseas. In: *18th Annual Conference of the association of US university directors of international agricultural programs*. Lincoln, NE.

Fraser, A.S. and Tinker, I. (eds.) (2004). *Developing power: How women transformed international development*. New York: Feminist Press, CUNY.

Gibbon, D. (2012). Chapter 5 methodological themes in farming systems. Research and implications for learning in higher education. In: I. Darnhofer, D. Gibbon and B. Dedieu, eds., *Farming systems research into the 21st century: The new dynamic*. Netherlands: Springer, pp. 95–119.

Harwood, J. (2012). *Europe's green revolution and ever since. The rise and fall of peasant-friendly plant breeding*. London and New York: Routledge, esp: The green revolution and its critics; pp. 115–137; reforming the revolution, 1970–2010, pp. 137–165.

Hillenbrand, E., Karim, N. Mohanraj, P. and Wu, D. (2015). *Measuring gender transformative change: A review of literature and promising practices*. CARE USA. Working Paper.

Interagency Gender Working Group (IGWG). (2017). *Gender integration continuum 2017*. United States Agency for International Development. Available at: www.igwg.org/wp-content/uploads/2017/05/FG_GendrIntegrContinuum.pdf [Accessed March 2018].

IRRI. (1985). *Women in rice farming*. Proceedings of a conference on Women in Rice Farming Systems, IRRI, 26–30 Sept. 1983. Los Baños, Philippines: IRRI and Aldershot: Gower.

IRRI. (1979). *Farm-level constraints to high rice yields in Asia: 1974–1977*. Los Baños, Philippines: IRRI.

Jackson, C. (2005). *Strengthening food policy through gender and intra-household analysis: Impact assessment of IFPRI's multi-country research*. Impact Assessment Discussion Paper 23. Washington, DC: IFPRI.

Jiggins, J. (1986). *Gender-related impacts and the work of the international agricultural research centers*. CGIAR Study Papers Published by the World Bank, nr 17.

Karlsson, S. and Russell, N. (2017). *Mainstreaming gender in CGIAR research: The added value of a network approach*. Gender Research Action Plan Brief No. 1. CGIAR Gender and Agriculture Research Network. Cali, Colombia. Plus Reviews 2017.

Kauck, D., Paruzzolo, S. and Schulte, J. (2010). *CGIAR gender scoping study*. Washington, DC: International Center for Research on Women (ICRW).

Klerkx, L., van Mierlo, B. and Leeuwis, C. (2012). Evolution of systems approaches to agricultural innovation: Concepts, analysis and interventions. In: I. Darnhofer, D. Gibbon and B. Dedieu, eds., *Farming systems research into the 21st century: The new dynamic*. Dordrecht: Springer, pp. 457–483.

Kortright, C. (2012). *C4 rice and hoping the sun can end hunger: Tales of plants, evolution, transgenics and crisis*. unpublished PhD Thesis University of California.

Mead, M. (1976). A comment on the role of women in agriculture. In: I. Tinker and M.B. Bramsen, eds., *Women in development. American association for the advancement of science (AAAS)*. Washington, DC: Overseas Development Council, pp. 9–13.

Moock, J.L. (1989). Rockefeller foundation social science fellowship program in agriculture. In: D. Groenfeldt and J. Moock, eds., *Social science perspectives on managing agricultural technology*. Sri Lanka, Colombo: International Irrigation Management Institute (IIMI), p. ix.

Moock, J.L. (ed.) (1986). *Understanding Africa's rural households and farming systems*. Boulder, CO: Westview Press.

Moser, C. (1989). Gender planning in the third world: Meeting practical and strategic gender needs. *World Development*, 17(11), pp. 1799–1825.

Mowbray, D. (1995). *From field to lab and back, women in rice farming systems*. Washington, DC: CGIAR, The World Bank.

Norman, D.W. (1978). Farming systems research to improve the livelihood of small farmers. *American Journal of Agricultural Economics*, 60(5), pp. 813–818. Available at: www.jstor.org/stable/1240097.

Norman, D.W., Simmons, E.B. and Hays, H.M. (1982). *Farming systems in the Nigerian Savanna: Research and strategies for development*. Boulder, CO: Westview Press.

Oasa, E.K. (1985). Farming systems research: A change in form but not in content. *Human Organization*, 44(3), pp. 219–227.

Overholt, C., Anderson, M.B., Cloud, K. and Austin, J. E. (eds.) (1985). *Gender roles in development projects: A case book*. West Hartford, CT: Kumarian Press.

Paris, T. (2000). *Bringing women from the margin to the mainstream of rice research and technology development: Strategies and lessons learned*. PhD dissertation, Faculty of Social Inquiry (Social Ecology), University of Western Sydney, Hawkesbury, Australia.

Perkins, J.H. (1997). *Geopolitics and the green revolution: Wheat, genes, and the cold war*. New York: Oxford University Press.

Petesch, P., Badstue, L. and Prain, G. (2018). *Gender norms, agency, and innovation in agriculture and natural resource management: The GENNOVATE methodology*, Mexico, DF: CIMMYT. Available at: http://gender.cgiar.org/wp-content/uploads/2018/02/GENNOVATE-Methodology_Feb2018_FINAL.pdf [Accessed March 2018].

Poats, S.V. (1991). *The role of gender in agricultural development*. Issues in Agriculture no 3. Washington, DC: CGIAR.

Poats, S.V. (1990). *Gender issues in the CGIAR system. Lessons and strategies from within*. Paper presented at the Consultative Group Meeting, May 21–25, The Hague, The Netherlands.

Poats, S.V., Schmink, M. and Spring, A. (eds.) (1988). *Gender issues in farming systems research and extension*. Boulder, CO: Westview Press.

PRGA Program. (2011). *Final report of the PRGA program 1997---2011 / Program on participatory research and gender analysis*. CIAT Working Document No. 220. Cali, Colombia: CGIAR, CIAT.

Quick, G.R. and Paris, T.R. (eds.) (1995). *Enhancing incomes of rural women through suitably engineered systems*. Proceedings of an International Workshop 10–13 May 1994. IRRI discussion paper series 8. Los Banos, Philippines: IRRI. Available at: http://irri.org/resources/publications/serials/item/enhancing-incomes-of-rural-women-through-suitably-engineered-systems?category_id=2.

Quisumbing, A., Meinzen-Dick, R., Raney, T.L., Croppenstedt, A., Behrman, J.A. and Peterman, A. (eds.) (2014). *Gender in agriculture. Closing the knowledge gap*. Dordrecht: Springer.

The Rockefeller Foundation (RF) and International Service for National Agricultural Research (ISNAR). (1985). *Report of a seminar women and agricultural technology: Relevance for research. Volume 1 – Analyses and conclusions.* Report from the CGIAR Inter-Center Seminar on Women and Agricultural Technology, Bellagio, Italy, 25–29 March 1985, The Hague, ISNAR; *Volume 2, experiences in international and national research.* (volume II was also distributed with a slight difference in numbering: as ISNAR (1985) The Users' Perspective in International and national agricultural research. A background Document Prepared for a CGIAR Inter-Center Seminar on Women and Agricultural Technology, March 25–29, 1985, Bellagio, Italy). Bellagio, Italy: ISNAR.

Russell, N., Karlsson, K., Ashby, J. and Mascarenhas, M. (2015). *Change in the making: Progress reports on CGIAR gender research.* Issue No. 1: Toward gender-equitable control over productive assets and resources. Cali, Colombia: CGIAR Gender and Agriculture Research Network, CGIAR Consortium Office and International Center for Tropical Agriculture (CIAT).

Smith, E. (2009). Imaginaries of development: The Rockefeller foundation and rice research. *Science as Culture*, 18(4), pp. 461–482.

Swaminathan, M.S. (1985). IRRI's research and training agenda. In: IRRI, ed., *Impact of science on rice.* Los Baňos, Philippines: IRRI, pp. 41–63.

Yudelman, M. (1976) The role of agriculture in integrated rural development projects: The experience of the world bank. *Sociologia Ruralis*, 16(4), pp. 308–325.

Zandstra, I. (ed.) (1989). *Appropriate technology for rural women: Lessons from China.* WIRFS Seminar report 7–17 May 1989. Los Baňos, Philippines: IRRI.

Zeidenstein, G. (1978). Including women in development efforts. *World Development*, 6(7–8), pp. 971–978.

4 How to do gender research? Feminist perspectives on gender research in agriculture

Ann R. Tickamyer and Kathleen Sexsmith

Introduction

Why do we need feminist approaches to research on gender and agriculture? After all, many physical scientists and some social scientists would claim that feminist approaches are scientifically unreliable. Common concerns we hear are that feminist researchers have abandoned objectivity; that the results produced by feminist methods are not generalizable because they are often based on "small N" qualitative studies of specific groups in particular places; that they are partial and incomplete because they only focus on and care about women; and that the feminist approach is a thinly veiled political platform for redistributing resources and power away from men in favor of women. Critiques and suspicions of feminist research abound. As a result, feminist researchers are vulnerable to marginalization by mainstream agricultural researchers, even though the feminist approach offers a robust framework for understanding pressing questions about sustainability and prosperity in the agricultural sector.

This chapter asserts the value of feminist epistemologies, theories and methods for research on gender and agriculture. It provides conceptual tools for researchers seeking to persuade doubtful physical scientists and others of the importance of a feminist approach. After all, gender is shaped by social, political, biological and ecological factors and, in turn, exerts its influence on them, irrespective of disciplinary claims to knowledge. Thus, feminist and other agricultural researchers must work together toward their common goals of understanding and promoting more socially and environmentally sustainable agricultural systems. The chapter is organized around three main sections. First, we explain the significant difference between research on "gender" and research on women and emphasize the importance of a relational approach. Next, we describe feminist epistemologies, methodologies and methods and explain distinctions between them. In the final section, we identify obstacles to implementing feminist methods in the field and illustrate how researchers (including many in this volume) have managed to mitigate or overcome those obstacles. In doing so, we respond to the claims and fears of more conservative traditions in the physical and in the social sciences.

This chapter makes several central claims. We emphasize the superior or "strong" objectivity of feminist methods (Harding, 1992), which require that researchers acknowledge their positionality, and integrate that reflexivity into their results in a way that improves rather than undermines their authenticity. We also highlight the fact that feminist research can and often does combine qualitative and quantitative methods to improve both the depth and breadth of findings. Another central claim of the chapter is that good feminist research is not reduceable to research on women but rather includes and responds to the concerns of men and masculinities. Moreover, we argue that a gender-transformative research agenda can make positive contributions to the lives of women, men, and communities. A gender-transformative approach seeks to challenge and overturn structural inequalities between women and men by addressing the locally specific attitudes, social norms, political-economic systems and practices that sustain them (see Hillenbrand et al., 2015). Gender-transformative programming often entails working and talking with both women and men about the harmfulness of gender discrimination and the intrinsic rights to which every individual is entitled.

The global agriculture and development community is moving (albeit slowly) toward the integration of gender analysis into their research agenda. Indeed, many of the authors in this volume are post-doctoral researchers with CGIAR centers worldwide, illustrating the global uptake by donors and researchers of a more comprehensive gender-integrated agenda. Greater attention to gender is also reflected in the comprehensive integration of gender equality concerns in the 2030 Development Agenda, commonly known as the Sustainable Development Goals (SDGs),[1] including attention to gender inequalities in agriculture (Sexsmith, forthcoming). Therefore, while social science research takes many forms, it has increasingly moved away from a hard positivist stance to recognize many of the principles articulated in a feminist perspective, although not always coming from a specifically acknowledged feminist standpoint.

This chapter emerged from the first author's experience co-developing and co-leading training sessions on the integration of gender into agricultural research for participants in the Gender Research Integrated Training (GRIT) seminar, a collaboration between Penn State University and the CGIAR Gender and Agricultural Research Network over the course of three years. Several of the authors of chapters in this book participated in these trainings. As such, the purpose of the chapter is to serve as a companion to the GRIT training, specifically by summarizing and sharing the content. Importantly, the chapter is not meant as a comprehensive overview of feminist methodological issues, which would lie beyond its scope. Instead, we identify certain core feminist research concerns and locate them in the context of research on gender and agriculture. For a comprehensive exploration of and guide to feminist research methodologies, see Hesse-Biber (2012) and Hesse-Biber and Yaiser (2004).

In accord with feminist scholarship, we find it important to acknowledge and reflect on our own positionality with respect to the writing process. Ann Tickamyer and Kathleen Sexsmith are both rural sociologists at Penn State affiliated with the Department of Women's, Gender, and Sexuality Studies. Together they bring significant experience conducting feminist research in domestic and international contexts and have observed firsthand research on agriculture that is gender-blind, that misidentifies gender as a biological difference, that assumes the experiences of female household heads represents the experiences of all women and other problematic approaches. They are motivated to write this chapter by a desire to see gender more thoroughly and meaningfully integrated into agricultural research through feminist methods and by the hope that such research results can be used to promote gender equality and women's empowerment in rural places.

A feminist research agenda, as discussed more below, has an explicitly political agenda. Feminist research is conducted to help achieve social justice goals for socially marginalized women, men, girls and boys. In this regard, we advocate that adherence to feminist research principles can contribute to the achievement of the aspects of those elements of the 2030 Development Agenda (the SDGs) that relate to gender equality and women's empowerment in agriculture. These include the promotion of household food security, women's access to land and productive resources, gender equality in education, alleviation of women's unpaid care work burden, women's empowerment and decent working conditions (Sexsmith, forthcoming). However, there are significant gaps in data availability for many of the sex-disaggregated and gender-related indicators that countries use to report and monitor their progress toward the SDGs (UN Women, 2018). Thus, this chapter calls for the application of rigorous feminist methods in the field, both quantitative and qualitative, in order to move the global community toward greater gender equality and women's empowerment in agriculture.

Researching women vs researching gender

Too often, research framed around gender concerns only addresses and involves women, without attention to men, masculinities and gender relations. Moreover, such studies often treat women as a homogenous group without attention to the intersectional factors that shape their different standpoints, situated knowledges and perspectives and desires related to gender transformation. In this section, we describe the scope of this problem, reflect on why and how it occurs and offer some suggestions for avoiding the "gender as women" research trap.

Unit of analysis

A core issue in any research design is the unit of analysis: who or what is the focus of the research and how will the information be acquired? As we have indicated, research on agriculture has too frequently assumed a male subject and has proceeded accordingly – collecting data from and for men with little regard for women's critical roles in agricultural work, including differences and similarities to men and how their participation in agriculture matters. Alternatively, the household (or sometimes the family) is assumed to be the basic unit of analysis without reference to sex or gender, divisions of labor and differences in both objective roles and subjective perceptions within that unit. Such an approach both assumes and reinforces male dominance.

Even when data have been generated separately by sex, all too often it still is not collected systematically and rigorously, nor does the disaggregation get carried through to analysis or reporting. Sex-disaggregated data, or data that is "cross-classified by sex," improves the researcher's ability to identify inequalities and conduct a gender analysis.[2] Reports will specify agricultural activities by families, households, agricultural position (e.g., farmer, waged worker) with no indication of who is doing what, how this varies by gender or the consequent differential impacts by gender. Additionally, even when gender is accounted for, distinctions are often ignored or obscured between very different groups of women and very different groups of men. For example, household studies that collect data from both male and female household heads frequently conflate these two groups with men and women more generally. It is unfortunately common for both male and female household heads to be assumed to represent any and all gender differences. Variation by other social locations such as age, marital status, ethnicity, legal status and others create different social relations, necessitating an intersectional approach to gender that is infrequently applied.

Since one of the purposes of feminist research is to bring women's presence and experience to the fore, promoting equality, empowerment and the gender transformations necessary to achieve these objectives, the ultimate goal is to conduct gender analysis, i.e., to understand the ways that women and men interact and relate in the conduct of daily life and their productive and reproductive activities. This may lie beyond the purview of some single pieces of research, but the ultimate goal is to be able to understand gender relations in agricultural, ecological and environmental systems and how these create power relations that undergird the social system. By gender relations, we mean "the specific sub-set of social relations uniting men and women as social groups in a particular community, including how power and access to and control over resources are distributed between the sexes."[3] Thus, focusing on women "versus" gender is misleading. Rather, the task is to understand gender structures and processes, the dynamics that drive gender relations, the gender division of labor, the hidden "costs" of gender roles and relations and the power differentials that may be both obvious and obscured.

This brings matters back to deciding on units of analysis and the critical issue of what sorts of data will address and answer the specific research questions at hand. There are times when the focus will be specifically on women or on men (since research in the agricultural sector focusing on gender non-conforming and trans identities is woefully lacking), other times when the focus is on certain groups of women or men (e.g., heads of households, single women, etc.) and still others when the issue is comparative and relational, necessitating studies of women and men and interactions between them. Ultimately, all of the above are necessary for a holistic understanding of agricultural systems, structure and practice but may not be feasible or desirable in any particular piece of research given common limitations on funding, time and scope. Thus, the researcher needs to decide which group can best provide the desired information and whether it answers their questions. In practice, this means that careful thought needs to go into data collection procedures. If all women are of interest, how will they be found and selected? The same question applies when only some categories of women or of men are of interest according to an intersectional analysis, with the caveat that the reasons for focus on specific groups and subgroups should be clearly understood and specified.

Ultimately, the effort must be on gender relations. The relational approach specifies that men and their gender roles and relations with women are a subject of research interest rather than see men as the de facto explanation for all facets of women's disempowerment and of gender inequalities. Rather, men's participation should be defined by how male social roles and practices are constructed and by the meaning of masculinities. This opens important new avenues for research that previously have been assumed or neglected. Are certain agricultural practices tied to definitions of manhood in the context under scrutiny? What happens to women, children and the elderly when men are not available to work? What happens when women attempt to assert their interests and rights in the face of male disinterest or opposition? How are these roles maintained, enforced and reproduced? A feminist approach to research directs understanding of both women and men as gendered subjects and ultimate scrutiny of gender as a system, whatever the focus of specific studies. Research on women and men is not the same as research on gender but comprises an integral part of understanding gender.

Feminist research

The core principles of feminist research require embracing a feminist epistemology in which the researcher interrogates both the participant and the researcher's subjectivity to determine how they influence the conduct and outcomes of research. Feminist epistemology asks: What is knowledge? How and what do we and can we know? It produces different answers than standard positivist ways of knowing. In particular, gender (as well as other social factors) shapes experience that informs understanding of how the world works, an epistemological proposition known as feminist standpoint theory (Harding, 1992). Standpoint theory posits that women and other socially marginalized groups have different and more complete knowledge of systems of power because they experience their truest effects. Dominant research practices tend to ignore, exclude or demean gendered knowledge to the detriment of both women and men, with resulting failures to fully explain or understand the issue at hand attributable to the partial knowledge of the researcher. Adopting a feminist approach to research entails situating the research in larger social worlds and locations.

There are numerous variations on feminist epistemology and research. However, basic to all of them is understanding that all knowledge is situated, a basic tenet of standpoint theory. What we know reflects the body and mind of the knower. To conduct rigorous research with a claim to be furthering knowledge, it is necessary to recognize the inability to claim some external objective stance free of one's social location and to embrace instead embodied knowledge as an additional and fundamental source of information and understanding of the social world. Thus, feminist epistemology assumes researcher subjectivity and builds on it. In Sandra Harding's (1992) term, it promotes "strong objectivity" via the rigorous application of feminist standpoint theory. Strong objectivity refers to the need for the researcher to explicitly acknowledge their positionality. By accounting for their partial and situated knowledge, the research process will produce a "more ethical and transparent result" (Naples, 2007). This approach thus also calls for collective research processes to produce more holistic accounts from the collaboration of multiple situated knowledges (Naples, 2007). Integral to the strong objectivity perspective is a call to action to put results toward social justice objectives (Naples, 2007).

Feminist epistemology, theory, methodology and methods

Feminist research rests on related but different components of feminist perspectives that are often conflated with each other. That is, in addition to feminist epistemology, feminist theory, methodology and methods are all part of the repertoire of feminist research. Feminist theory uses the assumptions behind feminist ways of knowing, i.e., epistemology as discussed above, to develop concepts and their relationships to explain women's and men's experience, the roles of gender roles, relations and practices in larger society and how these are constructed and reproduced.

Methodology refers to research design – how the research will be conducted along with the justification, underlying assumptions and rationale for this choice. Feminist methodology necessarily assumes elements of feminist epistemology and theory and accepts the underlying premises of situated knowledge, strong objectivity and reflexivity. It does not specify the use of particular tools for conducting the research. This is the province of methods, which, simply stated, are the specific procedures that are used to collect and analyze data. Thus, methodology may specify quantitative, qualitative or mixed-methods designs, whereas the term "methods" will specify the use of surveys, experiments, interviews, focus groups, etc. Fonow and Cook (2005, p. 2214) illustrate this point with a clever table that specifies "selected methods employed by feminist scholars." The table lists dozens of different methods alphabetically, ranging from action/participatory to visual analysis with both well-known and more obscure tools listed in between. All the standard approaches such as surveys, experiments, focus groups, interviews etc. are included. Many others could be added, limited only by knowledge and imagination.

This list belies any assumption that feminist research is restricted to a particular set of tools, despite a historical affinity between qualitative approaches and feminist research designs. Partly, this can be attributed to the greater ease of exploring subjectivity and gaining deep understanding of process using qualitative methods; partly, to rebel against the hegemonic social science stance of positivist objectivity via quantitative methods; and partly as a reflection on the path by which feminism and science have interacted and co-developed. However, there is nothing inherent to a particular method of gathering data that makes it feminist or not, and a basic prescription would be to stay flexible in choice of methods to suit the problem at hand. Thus, rigorous gender research may be found using any and all methods or combination of them. What is important is how the tools are used: "research on gender must be emancipatory, must have a theoretical basis, must acknowledge its political underpinnings, and must incorporate reflexivity about the influence of social position on the knowledge produced" (Williams, 2010, p. 204).

Principles for feminist research

The basic components of conducting feminist research include a mix of feminist epistemology and theory, account for practical issues and promote a clear political agenda of gender equality and women's empowerment. Elaborating the discussion above, a list of basic principles and practices that should be recognized and incorporated into feminist research on agriculture would include:

- Pursuit of new and neglected research questions.
- Reflexivity and situated knowledge.
- Intersectionality and interdisciplinarity.
- Rejection of simple binaries.
- Distinction between difference, diversity and deconstruction of gender and its components.
- Study of both women/femininities and men/masculinities and their construction and interactions.
- Emphasis on power dynamics, inequality, empowerment and gender transformation.

Each of these could be and largely has been the focus of a disquisition on its own. In the limited space available, we briefly highlight key issues on this list. As a package, they characterize a feminist approach to research that directs the conduct of holistic, responsible research on gender and agriculture.

The first element, pursuit of new and neglected research questions, is arguably the most fundamental, in that pursuing a feminist approach requires raising new and previously neglected questions that have serious bearing on the overarching topic under investigation. It is well known that the history of science is full of research findings that assume a male subject and a male perspective. Thus, questions of reproduction as well as production, which includes unpaid care and domestic work as well as paid labor, questions on the availability, uses and impacts of new technology and knowledge, questions on time budgets and time poverty and on gendered divisions of labor, to name just a few, have often suffered from the failure to consider women and gender relations and have benefited from scrutiny with a gender lens. New questions bring new understanding, as well as progress toward gender equality and women's empowerment.

Reflexivity and situated knowledge have already been discussed. The important point for conducting gender research is that these cannot be ignored or neglected but rather make up a fundamental approach to the research. To achieve best practices, they should be continuously scrutinized throughout the research process.

Intersectionality has been variously classified as a theory, an analytical lens, a method, a process, a structure and a practice, reflecting ongoing discussion and debate about its exact epistemological status. For purposes here, however, what is relevant is that it is a feminist perspective that asserts the primacy of "the relationships and interactions between multiple axes of identity and multiple dimensions of social organization – at the same time" (Dill and Zambrana, 2009, p. 4). In the words of Patricia Hill Collins (2015, p. 2) "race, class, gender, sexuality, ethnicity, nation, ability, and age operate not as unitary, mutually exclusive entities, but as reciprocally constructing phenomena that in turn shape complex social inequalities." To fulfill the mandate for acknowledging situated knowledge of the researcher and research participants, it is necessary to understand the complexities of the social forces that construct roles and identities, particularly how they intersect and interact. In conducting research, it is necessary to incorporate other salient social locations into the research design and methods.

One hallmark of a feminist perspective is the rejection of simple binaries, starting with (but not limited to) conflation of gender with biological sexual distinctions between male and female, according to which all social roles are expected to align. Feminists also reject simple gender binaries between women and men, promoting recognition of trans and gender-non-conforming identities. Unfortunately, however, research that moves beyond the women–men binary is woefully lacking in gender and agriculture studies. Moreover, there may be obstacles in practice to a deep exploration of these spectra, when only binary gender identification data is available, and when it is not acceptable to ask about gender identity in certain study contexts where gender identity and women's empowerment are politically charged issues. Nonetheless, the complexity of roles and relations constructed around gender can be explored to identify, elaborate and deconstruct simple difference, diversity and multiple layers of meaning (Ackerly and True, 2010). It is critical that however gender is initially identified, its meanings and forms in context are explored. Thus, sex-disaggregated data are a necessity, and within-gender classifications and distinctions must be pursued where safe for the researcher and their subjects to do so.

Finally, gender research is underpinned by the feminist agenda of seeking equality, empowering women and promoting gender transformation in pursuit of these goals. Gender research conducted from a feminist perspective aspires to be transformative with social justice for women and for men as the ultimate goal. This supports "strong objectivity," which entails a call to produce results that are useful to a feminist agenda (Harding, 1992). This does not mean that outcomes are specified or data are "cooked" to produce desired results that benefit women. Rather, it requires recognition of gender bias or gender blinders and emphasizes the need for proactive efforts to overcome them while pursuing the knowledge and understanding that may ultimately advance these goals.

Challenges to doing feminist research in agriculture

This section outlines some of the principal challenges to conducting feminist research on gender and agriculture. Indeed, the persistent lacuna of feminist research in this area is due not only to hegemonic gender-blind ideologies in this field of study but also to real and pervasive practical barriers to implementing a feminist approach. Below we discuss the processes of and obstacles to feminist research on gender and agriculture. We also provide examples from this book wherever possible, and otherwise from the extant literature, on how feminist researchers have cleverly overcome some of these obstacles.

Research design

The first step in designing research is to define the research question(s). The second step is then to determine the most appropriate research design to answer the question(s) given the resources and skills at hand. Research designs that integrate gender fall into one of three basic categories: quantitative, qualitative or mixed. One other common design – participatory research – may use elements of any of the above while enlisting the active collaboration of research subjects in the study design, data collection, analysis and interpretation. Participatory research also raises questions about data ownership, requiring that communities have a full and equal sense of ownership to researchers. This becomes particularly important in a rural development context where communities can and should be able to use data in order to support requests for financial or other support from governments and donors. It is beyond the scope of this chapter to provide exhaustive accounts of these designs. Chapters by Berman et al. and Doss in Quisumbing et al. (2014) are excellent sources of applications of participatory research methods in agriculture.

The basic difference between the three different research designs lies in whether data can be expressed numerically and in the processes used to generate data. Quantitative methods use data collection and analysis procedures that permit attaching numerical values to the data. Quantitative methods typically include experiments, surveys and censuses. Categories are predetermined and can be manipulated arithmetically, generally using statistical techniques. Properly designed and conducted, quantitative methods permit assessment of the extent and scope of social phenomena, often with the ability to generalize to larger populations. Quantitative designs in this volume are illustrated in Chapter 6 by Camacho-Villa et al. The researchers implemented a large survey of 873 farm households in Guatemala and used multivariate analysis to identify five types of maize-producing households.

While quantitative methods achieve breadth of analysis, qualitative methods provide depth and nuance often by focusing on smaller samples. Qualitative methods often seek to understand the meanings, responses and resistances to social inequalities and power hierarchies. Some common qualitative methods include focus groups, interviews (structured, semi-structured or unstructured), life histories and participant observation. They can usefully be combined with quantitative methods to provide a holistic, descriptive and analytical summary answer to a research question. In Chapter 7, Kawarazuka et al. illustrate a particularly creative deployment of qualitative techniques. They employ participatory video production, screening, discussion, photo taking and four focus group discussions to study how young Hmong women negotiate and exercise their agency in upland agricultural and livelihood practices. This study also illustrates a participatory approach as it engages villagers in the generation and interpretation of the data.

Thus, quantitative and qualitative methods answer different types of questions. For example, understanding how agricultural production tasks are allocated by gender and how much time is devoted to each task and by whom in a community or culture is readily determined quantitatively, usually using survey, census or other sources of official statistics techniques. Understanding why this gender division of labor exists, what meaning is attached to it by participants and what problems they experience will require qualitative methods, such as in-depth interviews and focus groups conducted among much smaller numbers of participants.

Mixed methods that combine both types of data collection are increasingly popular and recognized as providing both the depth and breadth that cannot be supplied by either quantitative or qualitative studies alone. Numerous different approaches for conducting mixed-methods research exist, but all share a focus on extending understanding through integration of results from different data collection techniques (Axinn and Pearce, 2006; Cresswell and Clark, 2007). A number of the studies in this volume illustrate fruitful use of mixed methods. For example, in Chapter 13, Biketti et al. investigate "Kinship Structures, Gender and Groundnut Productivity in Malawi" with both women and men respondents using a combination of focus groups, key informant interviews and surveys. In Chapter 12, Gumucio et al. combine survey and interview data to look at gender empowerment in climate-smart practices in Nicaragua tree farming. Chapter 5 by Leder and Sachs also applies mixed methods in an intersectional approach we describe further below.

Sex-disaggregated data

Research on gender requires sex-disaggregated data. It is disappointing that at this late date, many data sets still conflate sex and gender or simply do not make disaggregated data available, either because it was not initially collected, was not collected systematically or is not reported in results. There is no remedy when sex-disaggregated data simply does not exist, and often there are few options when it does but is not made available to researchers. This limits the usefulness of secondary data that otherwise might be helpful in addressing research questions. The best that can be done is to make sure that going forward, gender is integrated into any research on gender and agriculture, regardless of whether the emphasis is on biophysical or social systems.

Equally important is to make sure that the techniques used to collect these data generate reliable answers. If the issue is the gender division of labor among household members, it is not enough to ask one household member, whether a woman or a man, who does what. Answers may vary due to lack of knowledge, ideological blinders or perceptions of what is normative or desirable. Rather, each household member should be given the opportunity to report their own labor responsibilities and the amount of time spent on them if time-use data is being collected. Also, women and men should be interviewed separately to ensure that responses reflect the respondent's true understanding and are not overly influenced by a dominant partner who is listening in on the conversation. Many more examples of the necessity of tailoring data collection to the research issue and the social circumstances of the individuals in question can be found in Doss (2014).

Intersectionality

Intersectionality illustrates the point that data collection and analysis can be conducted via a variety of research designs using either qualitative or quantitative methods, as long as salient social locations are identified and taken into account. Qualitative research can "unpack" the meaning of gender for women in different circumstances, deconstructing layered norms and practices behind gender differences, but is limited in its generalizability beyond the cases at hand. Quantitative data analysts have sometimes associated multiplicative models testing interaction effects among variables using large survey samples or census data as a means of specifying intersectionality (McCall, 2005). While a useful technique for identifying patterns in aggregate that reflect intersectionality, such modeling exercises are limited by data availability and sample size and are only one approach among many.

A useful and successful example of intersectional, mixed-methods research is found in Leder and Sachs's piece on measuring women's empowerment in Nepal (this volume). The authors compare results from one version of the WEAI (Women's Empowerment in Agriculture Index), an instrument designed to quantitatively measure women's empowerment, with results from a small selection of women's life histories, enabling them to discover and explain contradictory results. They find that while the WEAI research instrument would classify women as "empowered," the life history approach uncovered several intersectional factors that revealed how women's empowerment (defined in terms of their decision-making rights) in fact depended on who else was in the household and how the woman related to them. For example, age, caste, marital status and family support were all factors that determined their decision-making power and authority within the household.

Studying men and masculinities

As we have mentioned, the tendency to conflate research on gender with research on women is pervasive in the literature on gender and agriculture and on gender and development more broadly. While a literature on masculine gender identity in farming has emerged over several decades in the context of the Global North (Bell, 2000; Peter et al., 2000; Barlett and Conger, 2004; Alston and Kent, 2008; Bell et al., 2015), such work on masculinities and how they shape goals, practices, and needs in agriculture in the Global South is curiously lacking. (Perhaps as illustrated by the absence of a chapter on men and masculinities in the present handbook, a collection of cutting-edge research in the field of gender and agriculture.)

Why are men and masculinities so difficult to study in the context of research on agriculture for development? Practically, it is difficult to obtain funding for research on men, given historic discursive associations of women with disempowerment and of men with dominance, greed and violence. In short, donors have little desire to fund work that risks the appearance or practice of reinforcing or exacerbating existing gender hierarchies. The concept of "hegemonic masculinity" (Connell, 1987; Campbell and Bell, 2000) has proven exceptionally useful in describing how social norms around gender behavior for men have served as a harmful ideology that reinforces and exacerbates patriarchal norms, practices and institutions in different iterations across societies and historical moments. Unfortunately, the ways in which hegemonic masculinities can damage and disempower men and boys themselves have rarely been addressed, in the agricultural sector and more broadly. An intersectional lens can help identify groups of boys and men who are harmed by hegemonic masculinities that idealize strength, risk-taking, financial success, sexual prowess and other characteristics that are difficult to achieve.

The consequences of excluding men from research on gender can in fact create hardships for women.[4] For example, if research focuses heavily on women, so too will any policy recommendations arising from the research, thus potentially imposing the burden for transformative change on women themselves. Moreover, a research focus on gender equality and women's empowerment that excludes men can provoke a backlash from men who feel their status is being threatened or that they are being left out of the benefits. Backlash could take the form of domestic violence, withdrawal or withholding of money and other important resources for livelihoods, gossip and other harmful effects.

Despite these difficulties, there are some examples of successful research on men and masculinities in the Global South. The International Men and Gender Equality Survey gathered data on men's practices and behaviors with respect to gender equality in eight countries. Over 10,000 participants were surveyed for the project between 2009 and 2012 in Brazil, Chile, Mexico, India, Rwanda, DRC, Bosnia and Croatia (Levtov et al., 2014). The authors constructed a "Gender Equitable Men" scale composed by coding statements around gender equity attitudes, perceptions and support and used the scale to conduct several forms of bivariate and multivariate regression. They were able to use this analysis to make specific and practical findings, including that a gender-equitable home and education on gender equity contributes to more equitable attitudes and behaviors among men (Levtov et al., 2014).

Conclusion

This chapter selectively considers key issues for conducting good research on gender and agriculture from a feminist perspective, focusing on methodological challenges that are commonly faced by researchers in the field. While it is not a comprehensive examination of these issues, the selection of topics is informed by our own research experience, from trainings and collaborations that arose from the GRIT workshops and from examples provided by this book and others. Additionally, we have shown that feminist methods for research on gender and agriculture are not the unreliable, invalid and politically threatening set of tools that some physical and natural scientists might fear. Rather they promote a reflective, rigorous application of mixed methods that can advance understanding of how gender relations are intertwined with agricultural systems, policies and practices. In the process, we embrace an explicit feminist agenda in a sincere effort to promote social justice initiatives for women and men and girls and boys with sound evidence. Thus, we advocate for a gender-transformative approach that incorporates men and addresses masculinities that can be harmful to both men and women. It is necessary to uncover important knowledge about gender norms and attitudes in order to work toward making them more gender equitable. This approach can contribute to the advancement of women's rights, which are human rights. It can also contribute to global progress toward the SDGs if better and more rigorous data collection, monitoring and analysis is accomplished using a feminist approach.

Notes

1 Retrieved December 17, 2018 from www.un.org/sustainabledevelopment/sustainable-development-goals/)
2 Definition cited in UN Women Training Centre Gender Equality Glossary. Retrieved on December 18, 2018 from https://trainingcentre.unwomen.org/mod/glossary/view.php?id=36&mode=letter&hook=S&sortkey=&sortorder=
3 UN Women Training Centre Gender Equality Glossary. Retrieved on December 17, 2018 from https://trainingcentre.unwomen.org/mod/glossary/view.php?id=36&mode=letter&hook=G&sortkey=&sortorder=&fullsearch=0&page=1
4 Thanks to Carolyn Sachs for making these points in her GRIT training presentation on June 5, 2018.

References

Ackerly, B. and True, J. (2010). Back to the future: Feminist theory, activism, and doing feminist research in an age of globalization. *Women's Studies International Forum*, 13, pp. 462–472.

Alston, M. and Kent, J. (2008). The big dry: The link between rural masculinities and poor health outcomes for farming men. *Journal of Sociology*, 44(2), pp. 133–147.

Axinn, W. and Pearce, L. (2006). *Mixed method data collection strategies*. New York: Cambridge University Press.

Barlett, P.F. and Conger, K.J. (2004). Three visions of masculine success on American farms. *Men and Masculinities*, 7(2), pp. 205–227.

Behrman, J., Meinzen-Dick, R. and Quisumbing, A. (2014). Understanding gender and culture in agriculture: The role of qualitative and quantitative approaches. In: Quisumbing, et al., eds., *Gender in agriculture: Closing the gender gap*. Dordrecht, Heidelberg, New York and London: Food and Agriculture Organization of the United Nations and Springer, pp. 31–53.

Bell, D. (2000). Farm boys and wild men: Rurality, masculinity, and homosexuality. *Rural Sociology*, 65(4), pp. 547–561.

Bell, S. E., Hullinger, A. and Brislen, L. (2015). Manipulated masculinities: Agribusiness, deskilling, and the rise of the businessman-farmer in the United States. *Rural Sociology*, 80(3), pp. 285–313.

Campbell, H. and Bell, M. (2000). The question of rural masculinities. *Rural Sociology*, 65(4), pp. 532–546.

Collins, P. (2015). Intersectionality's definitional dilemmas. *Annual Review of Sociology*, 41(August), pp. 1–20.

Connell, R. (1987). *Gender and power*. Sydney, Australia: Allen and Unwin.

Cresswell, J. and Plano Clark, V. (2007). *Designing and conducting mixed methods research*. Thousand Oaks, CA: Sage.

Dill, B. and Zambrana, R. (2009). Critical thinking about inequality: An emerging lens. In: B.T. Dill and R.E. Zambrana, eds., *Emerging intersections: Race, class, and gender in theory, policy, and practice*. New Brunswick, NJ: Rutgers University Press, pp. 1–21.

Doss, C. (2014). Data needs for gender analysis in agriculture. In, Quisumbing, et al., eds., *Gender in agriculture: Closing the gender gap*. Dordrecht, Heidelberg, New York and London: Food and Agriculture Organization of the United Nations and Springer, pp. 55–68.

Fonow, M. and Cook, J. (2005). Feminist methodology. *Signs*, 30, pp. 2211–2236.

Harding, S. (1992). Rethinking standpoint epistemology: What is "strong objectivity?" *The Centennial Review*, 36(3), pp. 437–470.

Hesse-Biber, S. (ed.) (2012). *Handbook of feminist research: Theory and Praxis*. 2nd ed. Los Angeles: Sage.

Hesse-Biber, S. and Yaiser, M. (eds.) (2004). *Feminist perspectives on social research*. New York: Oxford University Press.

Hillenbrand, E., Karim, N., Mohanraj, P. and Wu, D. (2015). *Measuring gender transformative change: A review of literature and promising practices*. CARE USA. Working Paper.

Levtov, R., Barker, G., Contreras-Urbina, M., Heilman, B. and Verma, R. (2014). Pathways to gender-equitable men: Findings from the international men and gender equality survey in eight countries. *Men and Masculinities*, 17(5), pp. 467–501.

McCall, L. (2005). The complexity of intersectionality. *Signs*, 30(3), pp. 1771–1800.

Naples, N. (2007). Strong objectivity. In: G. Ritzer, ed., *The Blackwell encyclopedia of sociology*. Wiley Online Library and Available at: https://onlinelibrary.wiley.com/doi/pdf/10.1002/9781405165518.wbeoss286 [Accessed 17 December 2018].

Peter, G., Bell, M., Jarnagin, S. and Bauer, D. (2000). Coming back across the fence: Masculinity and the transition to sustainable agriculture. *Rural Sociology*, 65(2), pp. 215–233.

Quisumbing, A., Meinzen-Dick, R., Raney, T., Croppenstedt, A., Behrman, J. and Peterman, A. (eds.) (2014). *Gender in agriculture: Closing the gender gap*. Dordrecht, Heidelberg, New York and London: Food and Agriculture Organization of the United Nations and Springer.

Sexsmith, K. (forthcoming). *Leveraging voluntary sustainability standards for gender equality and women's empowerment in agriculture: A guide based on the sustainable development goals for development organizations*. Winnipeg: International Institute for Sustainable Development.

UN Women. (2018). *Turning promises into action: Gender equality in the 2030 agenda for sustainable development*. *United Nations*. Available at: http://www.unwomen.org/-/media/headquarters/attachments/sections/library/publications/2018/sdg-report-gender-equality-in-the-2030-agenda-for-sustainable-development-2018-en.pdf?la=en&vs=4332 [Accessed 5 March 2019].

Williams, J. (2010). Doing feminist-demography. *International Journal of Social Research Methodology*, 13(3), pp. 197–210.

Part 3

Intersectionality

Gender, class and ethnicity in agricultural systems

5 Intersectionality at the gender–agriculture nexus

Relational life histories and additive sex-disaggregated indices

Stephanie Leder and Carolyn E. Sachs

Introduction

The gender–agriculture nexus is often understood in gender and development studies through differentiating women's and men's labor and decision-making roles, gendered perceptions and participation, e.g., in resource user groups. Gender research has widely recognized that "women" and "men" differ in regard to their age, land ownership, ethnicity and other social and economic factors. However, approaches to further differentiate sex-disaggregated data, both collected through quantitative and qualitative methods, are often only additive through multiple categories, which has not fully incorporated a thorough understanding of "intersectionality." Instead, we understand intersectionality as "challenging ideas of difference that rely on fixed identities" (Nightingale, 2011). We argue that intersectional theory has two purposes in regard to the gender–agriculture nexus: firstly, intersectional theory deconstructs boundaries of gender and other social differences and draws attention to how the boundaries of particular gender categories are shifting in both time and space. Secondly, it situates gender construction in specific sociocultural contexts produced by power relations within households and communities or the state. Hence, engaging with intersectional theory is a de-hierarchizing approach and examines power relations that produce gendered subjectivities in particular contexts. Furthermore, intersectionality is useful to break down a number of assumptions about female and male farmers.

Our objective is to critically examine the role of intersectionality in research on the gender–agriculture nexus and the limitations of solely sex-disaggregated data collection. For this purpose, we link and compare qualitatively collected interview data and relational life histories of eight women with their responses and results of the sex-disaggregated quantitative survey, the abbreviated version of the Women's Empowerment in Agriculture Index (A-WEAI). Based on a binary understanding of gender, the A-WEAI has proved a powerful tool to raise awareness on gender in agriculture in international development discourses. However, being non-intersectional, assumptions on women in opposition to men can generalize on complex intersectionalities. If indices are used to compare gender data and the level of empowerment nationally or internationally, donors' decisions to allocate project funds to promote gender in research and development projects could be influenced and based on incomplete data and neglectng local meanings of empowerment. In our case study, a highly poverty-driven region in the Far Western Nepal hills, women were shown as empowered in decision-making, group membership, access to resources, and access to credit according to the index. As part of a broader qualitative fieldwork with index respondents in the region, we conducted both the A-WEAI questionnaire and relational life histories, observations and focus group discussions with several female farmers to demonstrate the value and greater validity of an intersectional approach that shows different results than the A-WEAI.

In this chapter, we will first review the emergence and importance of intersectionality as a feminist theoretical intervention. Second, we will reflect on studies on the gender and agriculture nexus and how these relate to intersectionality. Third, with a case study, we will illustrate the importance of intersectional understandings in regard to group membership, labor burden, control over land and water resources and decision making at the household level. We will do so by reflecting on four themes: (i) caste, class and inter-household relations; (ii) age, intra-household position and household composition; (iii) migration, 'female-headed households' and group membership; (iv) time and seasonality.

We will conclude with how to move forward with intersectionality research on the gender- agriculture nexus.

Intersectionality

One feminist theoretical intervention that proves useful in moving forward our thinking about gender and agriculture is the evolving theoretical and political work on intersectionality. Intersectionality focuses on "the overlapping and conflicting dynamics of race, gender, ethnicity, class, sexuality, nation, and other inequalities" (Lykke, 2010). Intersectionality emerged from the confluence of scholarship on feminist scholarship and critical race studies in the U.S. with a focus on the legal discrimination against African American women. Kimberle Crenshaw (1989), drawing on a long history of scholarship and activism by African American women, used the term to show how black women were discriminated against for both their race and gender and that the two categories could not be analyzed separately but involved complex and dynamic social relations. Patricia Hill Collins (2015, p. 2) writes about the definitional dilemmas of intersectionality and describes, "The term intersectionality references the critical insight that race, class, gender, sexuality, ethnicity, nation, ability, and age operate not as unitary, mutually exclusive entities, but as reciprocally constructing phenomena that in turn shape complex social inequalities." Intersectionality has been taken up by feminist scholars, critical race scholars and many other disciplines as a theory, method and strategy to understand and change overlapping systems of oppression based on race, class, ethnicity, sexuality and gender. Rather than focus on these categories as individual identities and subjectivities, intersectional understandings must emphasize political and structural inequalities (Cho et al., 2013).

While intersectionality has been widely adopted by feminist scholars across a number of contexts, some scholars have raised concerns that we must use caution and vigilance in uncritically adopting this theory without considering the ontological, epistemological and methodological foundations of the concept (Rodó-de-Zárate and Baylina, 2018). Recently, scholars have also called for looking at cross-border relations in intersectionality rather than focusing on intersectionality only within a domestic setting or nation-state (Patil, 2012). Patil argues that

> we need to think of the multiple processes, at different scales, that contribute to the emergence of particular local dynamics having to do with gender. From the perspective of intersectionality, merely reframing these dynamics through the lens of domestic intersectionality will only perpetuate the reification of the local. . . . We need to recenter the notion that there are no locals and globals, only locals in relation to various global processes.

Mollett and Faria (2018) also agree that intersectionality approaches can travel across borders and use the term postcolonial intersectionality to analyze how development processes are tied to patriarchy and race. As we attempt to address issues of gender and agriculture, understanding how interlocking systems of class, race, gender and ethnicity intersect in cross-border dynamics is essential. Building on a range of feminist scholars who have explored intersectionality in relevance to water and forest management, agricultural labor or climate change (Harris, 2008; Arora-Jonsson, 2011; Nightingale, 2011; Rao, 2014; Leder, Clement and Karki, 2017), we revisit the importance of intersectionality in gender and (agricultural) development research as complementary to sex-disaggregated data collection methods.

The gender-and-agriculture nexus

Gender norms, roles and relations needs to be viewed as a highly contextualized, multidimensional processes of which diverse women and men themselves have differing perspectives. Hence, it is important to understand subjectivities and the respective influencing factors in specific contexts. Approaches to understanding gender as a relational and processual concept take social structures and agency into account. Such understandings are crucially underpinning concepts such as empowerment, which define the scope of much gender research and developmental practice.

Different from the often individualized gender and empowerment discourse based on women's decision-making positions, Guérin, Kumar and Agier (2013) point out that women are "not necessarily looking for autonomy and independence from men, but rather for respect within their own community" (also see Basu, 1995; Kabeer, 2001; Cornwall, 2007). During their empirical study on microfinance in Tamil Nadu, they found that many women did not initially understand the term "empowerment," but when they had it explained to them, they defined it in their own words as the ability "to solve problems" and "to manage suffering" (Guérin et al., 2013, p. 80). Some related empowerment to material constrains, while others related it to the freedom of mobility and being respected by their family. This shows that women perceive their empowerment not as emancipation from male domination but in relation to other women, their husbands, families and communities. Guérin et al. (2013, p. 76) stress the importance of women's relationships with one another: "even where there is solidarity between women, women having agency require or imply domination over other women."

Particularly in the South Asian context, the interaction of multiple factors influencing women's agency needs to be taken into account. These factors include the educational status and land ownership, the influence of which can be ambiguous and needs to be considered in particular contexts. Trommlerová et al. (2015) provide an overview of and add to studies on gender correlates and variables of empowerment. With household-level information and advanced econometric techniques, they measure the correlation between a number of factors with individuals' self-reported ability to induce changes in their lives at both communal and individual levels. Their statistical findings suggest that age, gender, marital status, nationality, economic activity, health, self-reported capabilities and communal empowerment determine empowerment at both levels, suggesting that these findings may help to develop targeted policies towards gender, age and other social divides. Despite the importance of pointing out the multiple interacting factors with the category "gender," quantitative and additive approaches like this miss further relational and processual perspectives that in-depth qualitative studies can provide. In the following, we show the relevance of qualitative studies and how particular assumptions can be debunked.

Gender and intersectionality in agricultural decision making

The assumption that women's participation in agricultural decision-making bodies and grassroots organizations is always empowering or leads to better political outcomes has been rejected by a range of scholars on the basis that the category of women needs to be unpacked – as women benefit differently and some women might benefit more than others (Zwarteveen and Neupane, 1996; Agarwal, 2001; Leder et al., 2017; Arora-Jonsson, 2013). Joshi (2014), based on a study of water inequalities in Darjeeling, states that women in positions of power are as affected as men by political coercion structured by gender, class, ethnicity and other social divides, which influence context-specific cultural norms. Women experience a structural and symbolic relation with environmental resources derived from a "universal patriarchy" (Molyneux, 2001), but this relationship is crosscut by ethnicity, class, color, race and religion and evolves spatially and temporarily, leading to varying experiences. "Individual needs and priorities take precedence in the lives of women in political positions" (Joshi, 2014, p. 252), leading them to be unable or unwilling to address the complexity of water injustices in politics and in terms of a water crisis. Joshi (2014) outlines challenges to a politics of solidarity among diverse groups of women with differing needs, challenges and individual priorities. She argues that simplifying complex realities depoliticizes social hierarchies and inequalities (p. 253). Her study demonstrates how politics, ethnicity, class and religion interfere with an expected solidarity among women.

She argues for viewing gender as relational identity and not as social difference. This means that a homogeneous perspective on women may exclude marginalized women (e.g., Dalits) and may not lead to the change expected.

Agricultural development interventions, such as climate and water security interventions in Nepal, have been found to use the term "women's empowerment" apolitically and technically framed, failing to understand complex intersectional inequalities (Leder, 2017; Leder et al., 2017). Interventions that included "women" in decision making led to some women benefiting more than others, as differences between women such as age, marital status, caste, remittance flow and land ownership disadvantaged some women to participate. Focusing on inter-household relations within communities, as well as intra-household relations, could help to update the understanding of women's empowerment in line with current feminist understandings of it as a processual, relational, and multi-dimensional concept (Leder et al., 2017).

Interestingly, Zwarteveen and Neupane (1996) brought evidence that non-participation in decision-making processes can actually give women a covertly powerful position in relation to their access to water resources. Their gender analysis of the Chhattis Mauja scheme in Nepal empirically shows that women, despite being excluded from the management of the scheme's organization, can have a strong agency and access to resources despite non-participation in the water user group. Such an understanding is further related to other concepts in gender and development research, such as Kandiyoti's (1988) "patriarchal bargains" and Sen's (1990) "cooperative conflicts."

Gender, intersectionality and agricultural labor

Workforce participation is hardly the determining factor of women's agency and well-being (Rao, 2014). Rao conducted a household survey and in-depth interviews in rural Tamil Nadu, which provided evidence that the nature and social valuation of women's work influences women's agency. Other factors were also identified as important influences, such as age and stage in life cycle (e.g., own and children's marital status), reproductive success (especially by the birth of a boy) and caste and economic status. Women's reproductive work (motherhood, educational status and marital status of a son) can have a strong impact on agency. For example, the social position of a young wife changes if she gives birth to a son by "gaining recognition . . . maintaining an image as a good woman, wife, and mother" (Rao, 2014, p. 11). This can lead to greater agency within the family on issues of education and health care of their children. Because masculine and feminine identities are constructed through different values, "men face social pressures to earn and provide, women face pressures to reproduce – in particular, to produce sons" (Rao, 2014, p. 12). Although this study was not conducted in the agricultural sector, it highlights the separate spheres of productive and reproductive space, in which decision making is relevant. Lundberg and Pollak (1993) also identify these separate spheres in a bargaining model used in marriages (Kandiyoti, 1988), which includes a non-cooperative equilibrium that reflects traditional gender roles. This understanding may be differently important in the agricultural context, where the productive and reproductive spheres are not as separated. For example, involvement in agriculture, depending on the type of work and access to land, can contribute to family nutrition. The type of work being done also influences the perception of it as drudgery or work burden, even if there is an increase in self-worth that results from contributing to the family's income (Rao, 2014, p. 4).

Conceptualizations of empowerment are relational (see examples in Figure 5.1), while operationalizations require measurable – usually individual-based – approaches to empowerment. Qualitative data often points out to the relational ties diverse women have with each other or within their households and communities as these are embedded in power relations (Kandiyoti, 1988; Rowlands, 1998; Jackson, 2013; Guérin et al., 2013; Rao, 2017). Quantitative approaches aim at giving clear visibility on women's role within agriculture across contexts (Longwe, 1995; Alsop and Heinsohn, 2005; Alkire et al., 2013).

Quantifying gender empowerment in agriculture through sex-disaggregated data

Intersectional approaches to decision making, group membership and labor may help diversify operationalizations of empowerment in the form of measuring individual decision making, the participation in groups and gendered labor divisions. Recognizing the lack of measurements and quantifications of empowerment in agriculture, Alkire et al. (2013) developed the Women's Empowerment in

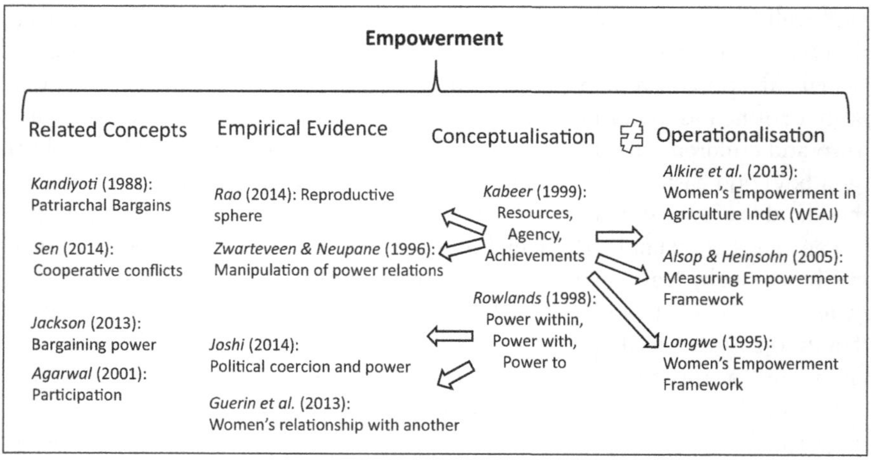

Figure 5.1 Relational conceptualizations and individualized operationalizations of gender and empowerment in agriculture

Source: own draft

Agriculture Index (WEAI). To quantify women's inclusion in the agricultural sector, the index measures five domains of women's empowerment relative to men: (1) decisions about agricultural production; (2) access to and decision-making power about productive resources; (3) control of use of income; (4) leadership in the community; and (5) time allocation. The WEAI questionnaire weighs answers on ten indicators according to the "adequate autonomy" (Alkire et al., 2013, p. 74) of women and produces a relative score that shows women's participation in decision making. If a woman reaches "adequacy" in 80% or more of the weighted indicators, she is empowered. The authors measure "agency," as defined by Kabeer (1999), while this score provides a situational and static quantification of empowerment from an economic perspective that may overshadow complex power relations within a particular context.

Feminist scholars have argued that Kabeer's definition of empowerment needs to be analyzed as a process and in its relation to resources and achievements (Leder et al., 2017; O'Hara and Clement, 2018). Particularly, social and human resources can play a great role in creating agency and may lead to different forms of empowerment that cannot be grasped by scoring decision making based on resources, production, income or representation in community groups. For example, not working or contributing to the household economically can be a choice and does not necessarily demonstrate disempowerment. Quantitative assessments on how time is spent (WEAI domain 5) could therefore be easily misinterpreted. Cornwall (2007) addresses empowerment and choice in the context of uncovering "gender myths":

> Seeing poor women as individuals who pursue entirely independent and goal-oriented strategies, as is often the case in discourses on "empowerment" and "choice" in development, is to deny the complexities of their relational ties and the contingencies of lived experience.
>
> (Cornwall, 2007, p. 158)

The WEAI labels women as either "disempowered" or "empowered," which can lead to their depiction as "victims" or "heroes." Furthermore, women who are the heads of households have scores that would indicate that they are "empowered" because they make all of the household decisions; however, they may have limited choices and feel strongly overburdened in their role as sole decision makers. Nevertheless, a simplified score may prove helpful in raising awareness of women's empowerment in international development discourses. To understand the cases in their respective contexts, however, in-depth gender analyses are necessary.

Methods

To critically examine the role of intersectionality in research on the gender–agriculture nexus, we compare sex-disaggregated quantitative survey data with evidence from qualitative social science methods collected from the same respondents. The study was set in two villages in Far West Nepal (Figure 5.2), a

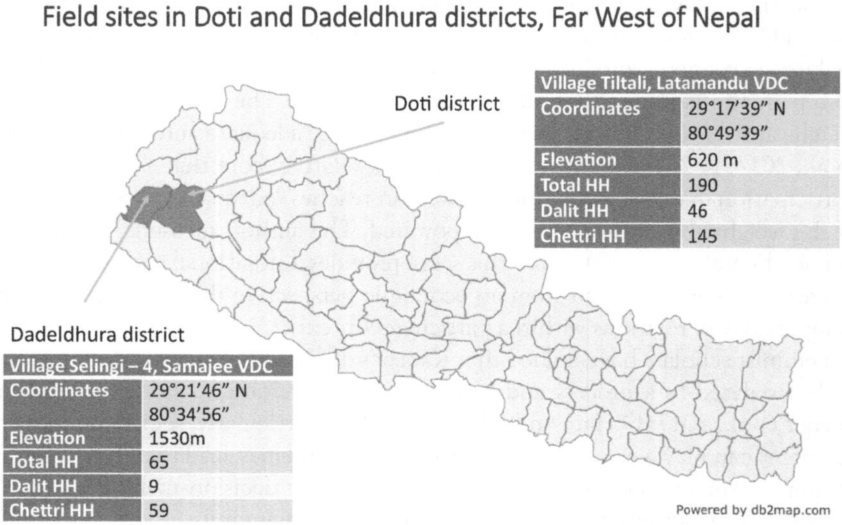

Figure 5.2 Location of two case study villages in Far West Nepal

region marked by a high rate of long-term male out-migration and a high dependence on remittances to manage household expenses. Specifically, we focused on eight households in particular. Based on in-depth interviews and life histories with women of the households, as well as participatory research methods such as transect walks, we aimed at gaining a deeper understanding about gender relations in agricultural practices and decision making. We conducted the A-WEAI for each household, and before reporting and discussing their individual results in visually guided focus group discussions with the respondents of each village, we collected, discussed and rated local notions of empowerment (Figure 5.3). For this purpose, women drew and wrote empowering factors they suggested on cards, and after discussing examples from their own experience, every participants received three stones to put on those factors they perceived most empowering for themselves. We also asked for each participants' self-assessment of A-WEAI dimensions after discussing with them their individual responses. Eventually, we linked the qualitatively collected data to the individual's questionnaire responses to the abbreviated version of the Women's Empowerment in Agriculture Index (A-WEAI). This offered a critical perspective on the role of intersectionality, as well as the limitations of sex-disaggregated data collection.

Results

Throughout our study, we identified a number of both conceptual and methodological challenges in regard to the validity of solely sex-disaggregated data and the value of intersectionality.

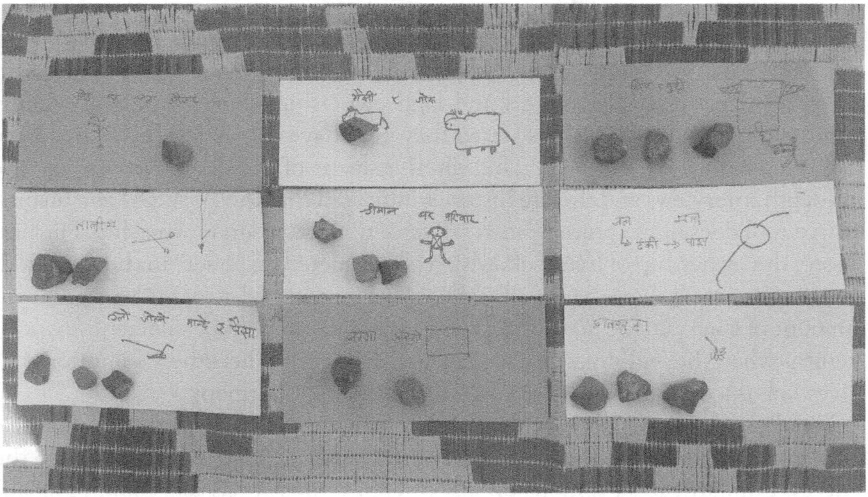

Figure 5.3 Participatory focus group discussion method on respondents' self-perception on empowering factors in agriculture. Five participants drew and wrote empowering factors on cards, and then selected the three most important factors for themselves by laying stones on the respective card

First, individual decision-making abilities are not necessarily considered empowering due to cultural norms and of viewing families as co-operative units. There was a tendency that Dalit women and female household heads were empowered in decision-making in their family according to the A-WEAI, but despite high decision-making power due to less restrictive cultural norms than in high castes or the absence of men. However, they are amongst the most marginalized in the community, with little access to natural resources or influence at the community level (Leder et al., 2017). Apart from that, powerful female leaders can also contribute to the disempowerment of other women, and hence, their success in decision making can reflect negative or "false" empowerment in developmental terms. This is particular common for elderly women with sufficient landholdings, earning sons or powerful husbands, who might overpower daughters-in-law.

Second, women's empowerment is highly relational to intra- and inter-household relations, which are not captured in the A-WEAI. Daughters-in-law of economically better-off Chettri households felt more empowered despite having less decision-making power in their joint families, as it is not culturally desired to be decision makers in their position. Political bonds to more powerful households within the village can be an advantage, too. This demonstrates the need to contextualize survey findings to avoid misinterpretations because of excluding inter- and intra-household power relations. This is even more important in contexts marked by male emigration and caste hierarchies. Relying on A-WEAI data to identify gender relations in agriculture across regions could misrepresent more marginalized women within joint households as empowered.

Third, diverse women attach diverse and different meanings to indicators that they intent to measure. This became particularly clear in regard to control over resources, such as land. In participatory focus group discussion on respondents' self-perception on empowering factors in agriculture (Figure 5.2), women differentiated between land access and land ownership, some considering family land as their land while others stated they only have the ownership if they have the document in their name. The self-assessment of A-WEAI dimensions, the in-depth interviews and the life histories indicated that the A-WEAI as a quantitative tool does not capture women's diverse notions of land ownership. Furthermore, the ownership of livestock, which is considered an "asset" in the A-WEAI, is not necessarily perceived so locally, as women stated to spend a significant amount of time cutting grass and taking care of them without sufficient financial returns when they sell livestock for income. This contradicts the assumption that livestock is a desirable asset per se that leads to empowerment.

Finally, empowering factors identified by women such as mental and labor support by their families and access to land ownership papers were not included in the survey. Similarly, local notions on empowered women (*shashaktikaran*in Nepali) included their ability to help others in agriculture and advise them on crops and cultivation methods: "They can tell us what to grow and how much to plant." These local perceptions on meaningful support through the sharing of knowledge and skills within households and communities are not captured in indices.

These results shed light on the limitation of solely focusing on sex-disaggregated data and the necessity of complementing those with qualitative studies on contextualized gender values and intersectionality. In the following, we highlight how intersectionality is a fundamental approach to apply when aiming to understand gender relations and empowerment in agriculture. We will examine this in regard to (i) caste, class and inter-household relations; (ii) age, intra-household position and household composition; (iii) migration, "female-headed households" and group membership; and (iv) time and seasonality.

There is a great need to contextualize survey findings of the A-WEAI (Figure 5.4) to avoid misinterpretations because of the exclusion of inter- and intra-household relations of power. This is even more important in contexts marked by male emigration and caste hierarchies, as this was highly prevalent in our two case study sites.

Caste, class and inter-household relations

There was a tendency that Dalit women and female household heads were empowered according to the A-WEAI, but despite high decision-making power due to less restrictive cultural norms than in high castes or the absence of men, they are often amongst the most marginalized within a community. In contrast, daughters-in-law of economically better-off Chettri households felt more empowered despite having less decision-making power in their families, as it is not culturally desired to be decision makers in their position. Hence, cultural norms and power relations in communities lead to diverse gendered relations.

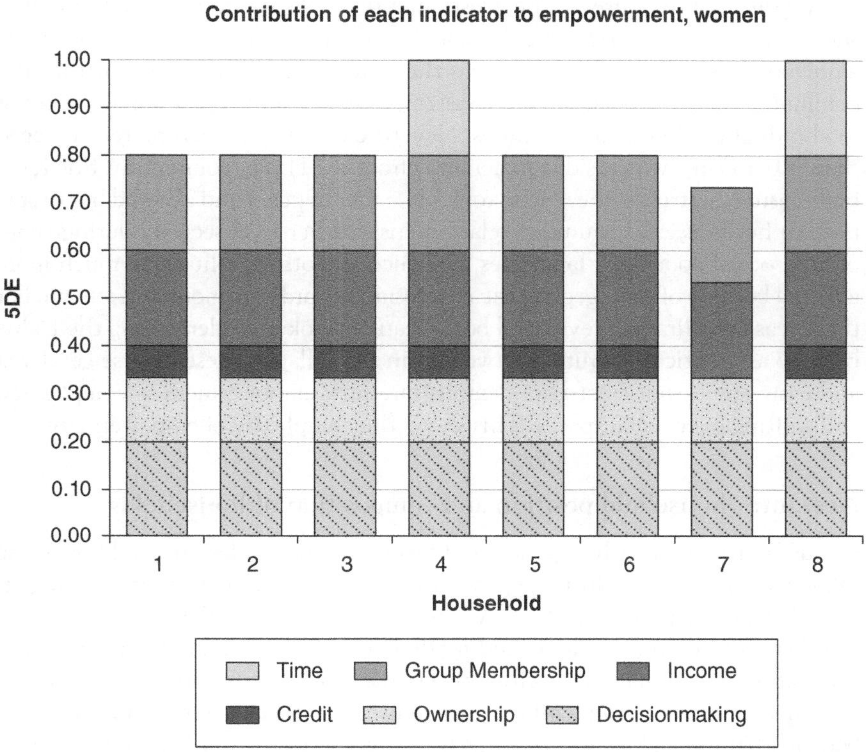

Figure 5.4 WEAI results of 8 women in households in our case study villages: Only household 7 was considered disempowered without reaching the threshold of 80%, as the female respondent was disempowered in both access to credit and time (labor burden)

The role of caste, class and inter-household relations became very apparent in the case of one of our landless Dalit research participants. According to the A-WEAI, Hira was empowered in all domains of the A-WEAI, except time. This was due to her having to work hard to ensure her two daughters and a son could attend school. Her husband has out-migrated to Malaysia and had not sent any remittances. In fact, she has not heard from him, and she felt burdened having to care alone for her children. Her eldest daughter lost her right hand's functionality during a recent accident, and she had no means to pay for medical bills. When we shared her A-WEAI results with her, she was surprised to see herself categorized as an empowered woman. As an agricultural wage laborer, she is dependent on landowning households for her family's food security. Going through the domains, she agreed that she was the one to make decisions and represent the household and that she feels empowered in these categories. However, as the tool does not contextualize the data and considers the respondents as individuals, and not as someone with relational ties to her household and the community, her responses to the questionnaire falsely lead to the conclusion that she is empowered in agriculture. Instead, her role within the community is highly ambiguous: as a Dalit, she belongs to the most marginalized groups within the community due to the Hindu caste system, as she cannot touch water sources or food of higher-Hindu-caste families. Nevertheless, she is much more privileged than other Dalits who reside further away from the Hindu community. She has a high-caste Chettri "mother," who took her under her care and allowed her to stay next to her house. This unique relation ensured her food security during times of monsoonal floods and landslides, by which the other Dalits were much more affected because of their geographical exposure towards the mountain stream. For these reasons, Hira has developed both as an outspoken leader among the Dalits but also as a critically scrutinized woman in the village. These diverse gendered and caste subjectivities of Hira demonstrate how empowerment is complicated and shifting over social spaces. This shows the complexity of empowerment.

Age, intra-household position and composition of households

Gender norms are highly dependent of intra-household relations, and how these influence participation in farmer meetings and resource user group meetings is not adequately captured in the A-WEAI in the context of Nepal. The relationship between daughters-in-law and mothers-in-law is particularly important, as the status of a woman rises with the marriage of her son, and even more so, the birth of (male) grandchildren. In return, her higher status can go to the benefit of the daughter-in-law, who has to comply to certain norms on manners and assigned (agricultural and reproductive) labor, which was before the domain of the mother-in-law. Hence, with a daughter-in-law, a mother-in-law has substantially less labor burden as she shares household responsibilities with her daughter-in-law.

In addition, daughters-in-law can be excluded from participating in farmer group meetings and agricultural trainings due to the presence of a mother-in-law. This exclusionary effect on a younger woman due to her mother-in-law participating and representing the household is not captured in the surveys, as only one female per household is interviewed. Furthermore, even if she would attend meetings, she could not participate to maintain cultural norms of respect towards her elder mother-in-law: "If she (the mother-in-law) is here, then I cannot say anything. She participates in trainings and meetings"(. This demonstrates how gender norms and roles in agriculture are shifting over time. While one woman within a household can attend meetings and agricultural trainings, another woman may be unable due to her lower position in the household, as well as time-consuming labor roles she is expected to fulfill. However, this does not necessarily mean that daughters-in-law feel disempowered, as another daughter-in-law shared as part of her life history: "They (in-laws) will tell me what to grow, where to grow. I don't feel like saying anything. They tell me what to do. I don't feel bad about it. That is fine. They know better." This is in line with Bourdieu's notion of "doxa," as gender (and age) inequalities within the household are culturally desired and not questioned.

Similarly, assumptions on the availability of loans for women has to be differentiated from actual control over credit. In several cases, the approval of elder female and male household members is required: "the wife cannot take a loan if the husband refuses to pay (back)." In other instances, females were requested to access loans to deliver the money to the head of the household.

Migration, "female-headed households" and group membership

Women with out-migrated husbands, who do not live with in-laws, (so-called female-headed households), are able to take decisions independently, but they might experience this as a burden rather than a form of empowerment. The A-WEAI scores indicated for one of our respondents in such a situation that she is empowered in decision making based on her position as household head without the presence of in-laws or her husband. However, the loneliness and worries about her son's health, how to pay back multiple loans for group memberships and her children's education, as well as her husband's security abroad are not captured by the A-WEAI. Furthermore, she lacks family support on financial and everyday matters such as child care. As phone calls to her husband are expensive, communication and mental support from him are very limited, and her maternal family lives far away. In fact, mental, financial and labor support by families was listed as empowering factors by women in focus group discussions; however, these relational aspects are not included in the survey.

These insights on women's positionality within households and families and perceptions about their role as "female household head" demonstrate how decision making and group membership measured by the A-WEAI can be considered a burden and that the measurement of the availability of loans does not necessarily translate into actual access and financial security. There have been a great number of prior development interventions in our study villages, which made every woman a member of at least one or several groups. However, group membership was in most cases token, as these groups were reportedly not functional.

Time and seasonality

The agricultural workload varies not only according to gender, class, caste, age and migration status of the household but also, and most importantly, with the seasons. Hence, it is important to reflect a reported lower workload in the A-WEAI as given at a fixed time of the year, and the data may vary a few weeks later, also across gender. For example, in focus group discussions around the gendered labor burden, perceptions on workload varied among women, as some stated to be burdened, and others felt the current workload was less in comparison to the end of the cropping season. Even more so, at days toward the beginning or the end of a cropping season, the workload of males can be particularly high as they conduct labor such as plowing and marketing the produce, which often falls under the male domain of work.

Conclusion and outlook

Our findings indicate that understanding cultural norms and power relations in communities is crucial for intervention planning, implementation, monitoring and evaluation. Women's age, caste, marital status, mental and labor support by families, remittance flow and land ownership shape their participation in decision-making processes on agriculture.

The A-WEAI survey results categorized women as empowered, but we found that women's decision-making abilities are highly dependent on the household composition and their position within the household. Female household heads tend to take more decisions due to their husband's absence; however, they often feel burdened and worried by the increased responsibilities of managing the household alone and in particular under climatic stress. Women living in joint families have among themselves power relations, depending on their position, with the mother-in-law having greater influence and authority as compared to the daughter-in-law.

As Sultana (2015) concluded on her women's perception of empowerment in Bangladesh, there is a need to "move beyond seeing women as victims or heroines, and engage with their everyday realities . . . to do more than give individual women economic opportunities . . . to tackle deeper-rooted structural constraints that perpetuate inequalities."

Gender indices like the A-WEAI data can raise awareness of the importance of gender in agriculture in international discourses. In order to identify specific gender-sensitive interventions, intersectional approaches are necessary to ensure the inclusion of diverse marginalized women.

In the context of agricultural research and development projects with community engagement, we identified a risk to reproduce or even exacerbate existing gendered inequalities and further marginalize female-headed households, Dalit women and young daughters-in-law through engaging primarily with already empowered community members. Therefore, the category of "women" needs to be further subdivided to ensure that women of different age and caste are represented, that is to ensure the participation of, for example, young mothers and daughters-in-law, time-constraint female household heads as well as Dalits.

Recommendations for development projects on the gender–agriculture nexus

In order to develop a better understanding of various power relations in specific communities, an intersectional feasibility study (see Table 5.1) can be conducted before the implementation of agricultural research or development project and the formation of farmer groups. We identified a need to go beyond technical feasibility surveys to also integrate assessment of social inequalities and in particular how multiple social markers intersect to shape access to natural resources within local communities. Since the influential households tend to dominate the decision-making process (even those who are not farming themselves), a participatory empowerment ranking exercise with the community members will help identify vulnerable and resource-weak households that should benefit from the project, particularly time-constraint female household heads, Dalits and young daughters-in-law. Including these households helps avoid reproducing existing power relations through project interventions and empowering those further who already hold powerful positions in the community.

We further recommend taking an intersectionality approach to women's empowerment by engaging project staff in planning and monitoring the interventions with diverse representatives of women in terms of age, caste, class and household composition to actively engage with sociocultural and economic divides amongst women (see Table 5.1). To encourage active participation of often marginalized groups such as female household heads, young female farmers and marginalized Dalit women, additional time and capacity building of project staff is necessary. For many project staff, we suspect that gender and caste are simply categories for reporting rather than markers of social inequalities that should be addressed in diverse ways to ensure wider benefits from interventions. Efforts must be made to actively engage representatives in the discussion process on new interventions, e.g., seed distribution or water tap installation, to cater to their diverse needs. Therefore, the local staff must be encouraged through participatory intervention processes to use the community's diverse knowledge for more inclusive approaches.

Agricultural interventions and trainings need to develop beneficiary categories appropriate at the village level, for example with at least three intersecting categories at the gender- caste/class – age nexus (= 8 subgroups) and differently approach different caste/class representatives of:

- Young men (sensitizing for gender issues, involving in programs as they otherwise become future migrants)
- Young daughters-in-law (child care as incentives)
- Female-headed households (adjusting timings and reducing work burden)

Recommendations for research on the gender–agriculture nexus

To examine the degree to which an intersectionality lens is applied in gender–agriculture nexus studies and development practice, we developed a simple tool named the "intersectionality grid" (Table 5.1). We plot common dimensions of social differentiation against sections of a research process (theory/approach, literature review/references, data collection/implementation, data analysis/monitoring, conclusion and recommendations) in which these could be addressed. This grid enables us to identify which dimensions of social differentiation are addressed and whether this is done in an additive (weak) or comprehensive (strong) manner. As some markers of social differentiation are stronger than others, and the type of marker varies as well (e.g., caste is mostly relevant in Hindu

Table 5.1 Intersectionality grid to review the degree of intersectionality applied in research and development practice

		Theory/ Approach	*Literature Review/ References*	*Data Collection/ Implementation*	*Data Analysis/ Monitoring*	*Conclusion and Recommendations*
Socioeconomic dimensions of differentiation	Gender					
	Age					
	Class					
	Religion					
	Caste					
	Family composition					
	Family Position					
Environmental dimensions	Rurality/Urban					
	Landscapes					

communities in South Asia), this grid just gives a first impression on which aspects can be covered. The degree of relationality explored in a study can be in depth by focusing only on the intersectionality between two markers. Hence it is important to review both the extent and quality of intersectional analysis or approach applied.

One can identified four degrees of how intersectionality is conceptionally developed and applied in studies:

1 **No intersectionality:** Gender is the only category of difference and findings, are not linked to other possible factors. This runs the risk of essentializing and mispresenting gender as binary, homogeneous categories of women vs. men.
2 **Weak, additive intersectionality:** Gender differences are linked to some other social differences such as age or class, and gender sub-categories are developed.
3 **Strong, comprehensive intersectionality:** Gender constructions and power relations are analyzed and situated in specific sociocultural contexts, which results in a comprehensive intersectional gender analysis.
4 **Full intersectionality:** The production of diverse and gendered subjectivities is explored, notions of gender are embedded in power relations within households and communities, or the state.

Acknowledgments

The authors would like to thank Floriane Clement, Corey O'Hara Ram Bastakoti and Emma Karki for their valuable insights on the empirical data and fieldwork support.

The research was partially funded by the DFID program BRACED (Building Resilience and Adaptation to Climate Extremes and Disasters) as part of the Anukulan project "Developing climate resilient livelihoods for local communities through public–private partnership for 500,000 poor people in western Nepal that suffer from climate extremes and disasters." Further, we would like to thank iDE and local staff for supporting our fieldwork.

References

Agarwal, B. (2001). Participatory exclusions, community forestry, and gender: An analysis for South Asia and a conceptual framework. *World Development*, 29(10), pp. 1623–1648.

Alkire, S., Meinzen-Dick, R., Peterman, A., Quisumbing, A., Seymour, G. and Vaz, A. (2013). The women's empowerment in agriculture index. *World Development*, 52, pp. 71–91.

Alsop, R. and Heinsohn, N. (2005). *Measuring empowerment in practice: Structuring analysis and framing indicators*. World Bank Policy Research Working Paper 3510.

Arora-Jonsson, S. (2013). *Gender, development and environmental governance: Theorizing connections*. New York: Routledge.

Arora-Jonsson, S. (2011). Virtue and vulnerability: Discourses on women, gender and climate change. *Global Environmental Change*, 21(2), pp. 744–751.

Basu, A. (1995). *The Challenge of Local Feminisms: Women's Movements in Global Perspectives*. Boulder, CO: Westview Press.

Cho, S., Crenshaw, K. and McCall, L. 2013. Toward a field of intersectionality studies: Theory, applications, and praxis. *Signs*, 38(4), pp. 785–810.

Collins, P. (2015). Intersectionalities definitional dilemmas. *Annual Review of Sociology*, 41, pp. 1–20.

Cornwall, A. (2007). Revisiting the "Gender" agenda. *IDS Bulletin*, 38(2), 69–78.

Crenshaw, K. (1989). Demarginalizing the intersection of race and sex: A black feminist critique of antidiscrimination doctrine. *University of Chicago Legal Form*, 1(1), pp. 139–167.

Edström, J. (2014). The male order development encounter. *IDS Bulletin*, 45(1), pp. 111–123.

Elmhirst, R. (2011). Introducing new feminist political ecologies. *Geoforum*, 42(2), pp. 129–132.

Guérin, I., Kumar, S. and Agier, I. (2013). Women's empowerment: Power to act or power over other women? Lessons from Indian microfinance. *Oxford Development Studies*, 41(suppl), pp. S76–S94.

Harris, L. (2008). Water rich, resource poor: Intersections of gender, poverty, and vulnerability in newly irrigated areas of Southeastern Turkey. *World Development*, 36(12), pp. 2643–2662.

Harris, L. (2006). Irrigation, gender and social geographies of the waterscapes of southeastern Anatolia. *Environment and Planning*, 24, pp. 187–213.

Jackson, C. (2013). Cooperative conflicts and gender relations: Experimental evidence from southeast Uganda. *Feminist Economics*, 19(4), pp. 25–47.

Joshi, D. (2014). Feminist solidarity? Women's engagement in politics and the implications for water management in the Darjeeling Himalaya. *Mountain Research and Development*, 34(3), pp. 243–254.

Kabeer, N. (2001). Gender equality and women's empowerment: A critical analysis of the third millennium development goal. *Gender and Development*, 13(1), pp. 13–24.

Kabeer, N. (1999). Resources, agency, achievements: Reflections on the measurement of women's empowerment. *Development and Change*, 30, pp. 435–464.

Kandiyoti, D. (1988). Bargaining with patriarchy. *Gender and Society*, 2(3), pp. 274–290.

Leder, S. (2017). Gender issues in water and sanitation programmes: Lessons from India. A book review. *SAWAS South Asian Water Studies*, 5(4), pp. 59–62.

Leder, S., Clement, F. and Karki, E. (2017). Reframing women's empowerment in water security programmes in Western Nepal. *Gender & Development*, 25(2), pp. 235–251.

Longwe, S. (1995). Gender awareness: The missing element in the third world development program. In: C. March and T. Wallace, eds., *Changing perception: New writings on gender and development*. Oxford: Oxfam.

Lundberg, S. and Pollak, R. (1993). Separate spheres bargaining and the marriage market. *Journal of Political Economy*, 101(6), pp. 988–1010.

Lykke, N. (2010). *Feminist studies: A guide to intersectional theory, methodology and writing*. New York: Routledge.

Mollett, S. and Faria, C. (2018). The spatialities of intersectional thinking: Fashioning feminist geographic futures. *Gender, Place & Culture*, 25(4), pp. 565–577.

Molyneux, M. (2001). *Women's movements in international perspective: Latin America and beyond*. New York: Palgrave.

Nightingale, Andrea. (2011). Bounding difference: Intersectionality and the material production of gender, caste, class and environment in Nepal. *Geoforum*, 42(2), pp. 153–162.

O'Hara, C. and Clement, F. (2018). Power as agency: A critical reflection on the measurement of women's empowerment in the development sector. *World Development*, 106, pp. 111–123.

Patil, V. (2013) From patriarchy to intersectionality: A transnational feminist assessment of how far we've really come. *Signs*, 38(4), pp. 847–867.

Rao, N. (2017). Assets, agency and legitimacy: Towards a relational understanding of gender equality policy and practice. *World Development*, 95, pp. 43–55.

Rao, N. (2014). Caste, kinship, and life course: Rethinking women's work and agency in rural south India. *Feminist Economics*, 20(3), pp. 78–102.

Rodó-de-Zárate, M. and Baylina, M. (2018). Intersectionality in feminist geographies. *Gender, Place & Culture*, 25(4), pp. 547–553.

Rowlands, J. (1998). A word of the times, but what does it mean? Empowerment in the discourse and practice of development. In: H. Afshar, ed., *Women and empowerment: Illustrations from the third world*. London: Macmillan, pp. 11–34.

Sen, Amartya K. (1990). Gender and cooperative conflicts. In: I. Tinker, ed., *Persistent inequalities: Women and world development*. Oxford: Oxford University Press, pp. 123–149.

Sultana, F. (2015). Rethinking community and participation in water governance. In: A. Coles, L. Gray and J. Momsen, eds., *The Routledge handbook of gender and development*. New York: Routledge, pp. 261–271.

Trommlerova, S., Klasen, S. and Lebmann, O. (2015). Determinants of empowerment in capability based poverty approach: Evidence from the Gambia. *World Development*, 66, pp. 1–15.

Zwarteveen, M. and Neupane, N. (1996). *Free-riders or victims: Women's nonparticipation in irrigation management in Nepal's Chhattis Mauja irrigation scheme*. Research Report 7. Colombo: International Water Management Institute.

6 Diversity of small-scale maize farmers in the Western Highlands of Guatemala

Integrating gender into farm typologies

Tania Carolina Camacho-Villa, Luis Barba-Escoto, Juan Burgueño-Ferreira, Ann R. Tickamyer, Leland Glenna and Santiago López-Ridaura

Introduction

Smallholder farmers manage 40% to 50% of the world's farmland and produce more than half of the world's food. However, there is little understanding of where those farmers are located, how they use their land and whether their agricultural production is subsistence or market oriented (Samberg et al., 2016). Knowing more about these smallholders can enhance knowledge of what they need to become more productive and to improve their general well-being. Such information could be useful for policy makers, research organizations like the CGIAR system and research universities, extension organizations and private firms that supply inputs. With better knowledge of the diversity of farmers they serve, each of these entities could better target knowledge and technology delivery in development interventions. We contribute to efforts to understand smallholder heterogeneity by developing a typology of smallholder farmers in the Western Highlands of Guatemala (WHG) that adds social dimensions to agronomic and farm management dimensions. The WHG represents 18% of that nation's land. Its mountain landscape creates several microenvironments due to a broad range of altitudes (1,450–4,220 m above sea level), temperatures (average 15–22 °C) and annual rainfall (1,000–3,500 mm year^{-1}) (Gálvez, 2000). This creates a variety of ecosystems in which different types of maize were developed, making this region one of the world's most important areas of maize domestication and diversification (Steinberg and Taylor, 2009). This environmental diversity also corresponds to differences in agricultural and maize productivity (MAGA et al., 2011).

Small-scale farmers in Guatemala are critical for food security, especially related to maize production-consumption. Farmers owning less than 7 ha of land represent 53% of the farmers growing maize in the country. Over two-thirds of the maize they produce is used for self-consumption that lasts for approximately eight months (USAID/Guatemala, 2009). These subsistence farmers are among the least food secure (Taylor et al., 2006) in a country where only 19% of inhabitants are food secure (Instituto Nacional de Estadítica, 2012). Food insecurity is rooted in inequality (Gobierno de Guatemala, 2005; Guardiola et al., 2006) linked to ethnicity, land ownership, health, education, gender and age (Bruni et al., 2009). Small-scale farmers from the Western Highlands with indigenous origins, as well as women, are consistently identified as those who suffer most from the country's high levels of inequality and food insecurity. Although these differences are recognized, the nature of small-scale farmers' diversity is understudied (Smith, 1989). Development interventions to improve their food security must take into account how this variation corresponds to diverse maize production-consumption strategies.

Farm typologies have been proposed as a means to understand and formalize farm household diversity and complexity since the 1970s and 1980s (Cochet, 2015). Farming Systems Research has been at the forefront of this work, identifying the structural and functional dimensions of diverse farming systems (see Tittonell et al., 2010; Alvarez et al., 2014; Berre et al., 2016). Social scientists have also highlighted the importance of heterogeneity by demonstrating the links between social identities and technology adoption (Bidogeza et al., 2009).

There have been efforts to create typologies of Guatemala farmers at different levels. A national-level typology divides farmers into (i) commercial large scale, (ii) commercial small scale, (iii) subsistence and (iv) infra-subsistence (Gobierno de Guatemala, 2012). This typology is premised on the idea that subsistence farmers with landholdings between 0.3 and 0.7 ha can supply their needs and are the main suppliers of staple crops in rural markets. Infra-subsistence farmers own less than 0.3 hectare and are not self-sufficient. Applying this typology at the regional level, MAGA et al. (2011) found that 65.6% of farmers from the Western Highlands belong to the infra-subsistence group and 28.3% to subsistence farmers. They also found that these two types of farmers own only 43.5% of the arable land but account for 93.9% of farmers in this region. This large percentage justifies the effort to document diverse structural, functional and social dimensions of their farms.

We build on previous efforts by incorporating social dimensions along with technical aspects and resource endowments. We construct these typologies from a multivariate analysis that integrates structural, functional and social characteristics of 873 farm households (FHH) distributed in the departments of Quetzaltenango, Totonicapan, Huehuetenango and Quiche. The following section describes historical trends of maize production and consumption in the Western Highlands.

Historical perspective of maize smallholder heterogeneity

During Pre-Columbian times, a slash-and-burn *milpa* system was the primary form of food provisioning. However, Wilken (1971) explains that there were also more intensive, sophisticated systems like terraces and *tablones* for producing native vegetables. During this period, production of maize and other food crops occurred mainly in the highlands in a self-subsistence scheme that included tributes to the local lords (van Etten, 2006).

The Spanish Colonial period brought changes, although the focus was still for local consumption. The introduction of new crops (like wheat) and domestic animals (like sheep) changed the regional landscapes, but unevenly. Farming was directed at self-consumption with an added burden of double tributes for the Spanish Crown and the Church (Elliott, 1989). Although renting became a common form of land dispossession, indigenous people from the highlands were able to maintain control over their land and labor (Mc Creery, 1989) and continued the highland/lowland maize production scheme.

In the 19th century, land conflicts (from population growth and the privatization of collective indigenous land) and the introduction of coffee in 1873 promoted temporary migration to coffee plantations through forced labor (Elliott, 1989). The dispossession of indigenous land helped lowland plantation owners buy highland plots to ensure that they had indigenous labor for their coffee and sugar plantations (McCreery, 1989). This led to the creation of three zones of crop specialization (van Etten, 2006). The heavily populated Totonicapan zone provided goods through trade. The central Quetzaltenango zone and southern Quiche specialized in vegetable and basic food production. Huehuetenango, northern San Marcos and northern Quiche supplied labor through seasonal migration (van Etten and Fuentes, 2004).

During the 20th century, these three zones remained distinctive regions in terms of commercial and local production characteristics. Smith (1989) uses this regionalization to describe the different survival strategies of petty commodity producers. Farmers located in the core zone of the Totonicapan Department specialize in agricultural and non-agricultural goods for markets. Although these farmers have the smallest holdings, they are the most productive agricultural systems, satisfying their maize needs in ten months. Farmers from the Central zone are distributed around four marketing systems representing the administrative towns of Quetzaltenango, Sololá, Chimaltenango and Santa Cruz Quiché. These farmers produce food for the market, even though they are rarely self-sufficient (having maize for 11 months). Therefore, they complement their production with seasonal work on plantations. Farmers from the peripheral zone, comprising four marketing systems crossing the north and the west region and covering North Quiche, Huehuetenango and North San Marcos, complement subsistence food production (only producing maize for eight months) with handicrafts, petty trade and seasonal labor force for plantations).

In the 1990s, Guatemala ceased being maize self-sufficient (van Etten and Fuentes, 2004). Currently the country imports maize to support one-third of its domestic needs. The Western Highlands is one area where maize production has failed to keep pace with national needs (Isakson, 2014), but maize continues to be a staple crop for the regional population with a per-capita consumption of 318 grams per day (Fuentes-López et al., 2005). The dependency on maize from other parts of the country and from Southern Mexico creates additional risks to food security (Hellin et al., 2017).

Several trends shape the current agricultural landscape of the Western Highlands. First, since the 1970s, the production of non-traditional export crops, along with a reallocation of land and labor, has fostered food deserts in the region (Isakson, 2014; Webb et al., 2016). Second, the political-economic shift toward trade agreements like the Central America Free Trade Agreement (CAFTA) and the decline in governmental infrastructure that supported maize production between 1960 and 1980 (van Etten and Fuentes, 2004; Sigüenza Ramírez et al., 2010) has exacerbated the problem. Third, the Civil War that lasted for nearly four decades involved the destruction of food systems as a standard military practice (Isakson, 2014). The fourth factor has been the extreme climate phenomena, especially a decline in rainfall and a rise in temperatures, which especially affects small-scale subsistence farmers (USAID/Guatemala, 2009). Finally, the fifth set of factors is transnational migration and remittances, which have been connected to a reduction of maize land (Moran-Taylor and Taylor, 2010). These trends, together with historical processes, explain the heterogeneity of farm types.

Methods

Two approaches generate farm typologies. The first, farm systems theory, considers the structural and functional components of farm households (FHH). Tittonell et al. (2005) studied soil fertility management strategies and developed a systems approach that incorporates both resource endowment (the structural) and resource allocation patterns (the functional) to explain smallholder farm heterogeneity. In Guatemala, several studies document farm typologies based on these structural and functional dimensions, using variables like land size, land tenure, self-consumption, market access, crop diversification, livelihood diversification, hired labor and income distribution (Fuentes-López et al., 2005; Taylor et al., 2006; Bellow et al., 2008; de Janvry and Sadoulet, 2010; Gobierno de Guatemala, 2012).

The second approach incorporates socioeconomic variables to identify farm differences. Although Meinzen-Dick et al. (2012) show the importance of incorporating social factors to identify farm differences, these variables are not commonly used to construct farm types (for exceptions, see Bidogeza et al., 2009; Baral and Bardhan, 2016). Socioeconomic attributes are especially relevant in the context of development projects in the WHG focused on reducing food insecurity and malnutrition, since those problems are rooted in inequality processes related to gender, age and ethnicity.

We combine structural, functional and socioeconomic factors in our analysis of farm differences (see Table 6.2). The structural variables include land availability, land pressure and the presence of animals. The functional variables include on-farm activities, such as cropping systems, and off-farm activities. The socioeconomic variables include gender, ethnicity and age. Annex I shows the variables used.

Data collection

Data were collected as part of the project Feed the Future Buena Milpa (www.cimmyt.org/project-profile/buena-milpa/). The project's goal is to reduce food insecurity

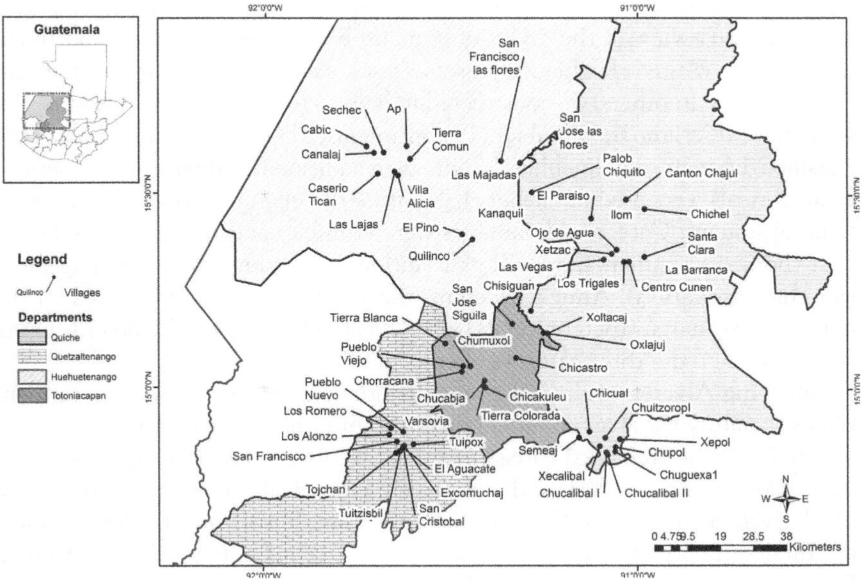

Figure 6.1 Map showing the communities that were surveyed in the departments of Quiche, Totonicapan, Quetzalenango and Huehuetenango

and malnutrition by fostering sustainable, resilient, and innovative maize-based farming systems in the GWH. The survey was conducted in summer of 2015 with 873 farm households cultivating maize in 59 communities of 16 municipalities. The number of surveys at the departmental level was: Totonicapán (202), Quiche (320), Quetzaltenango (152) and Huehuetenango (199) (Figure 6.1). Criteria to select communities were: location within the targeted municipalities in the Buena Milpa project, the active work of project partners within the communities and their distribution within four selected watersheds that cover farming systems at different altitudes. Since the aims of Buena Milpa are implementing a sustainable intensification strategy, the communities selected cover the agricultural and biological diversity in the WHG and are sensitive in capturing marginalized ethnic groups. For household selection in each community, enumerators walked radial transects to survey household members after they were informed of the goals of the survey and agreed to participate. Confidentiality was ensured by substituting an ID number for names. The survey had 139 questions including five project themes (milpa-maize germplasm improvement, natural resource conservation in farming system, farming systems diversification, agricultural innovation systems and social inclusion).

Multivariate analysis and cluster definition

The data analysis proceeded in three steps: (i) variable selection, (ii) multivariate analysis, dimension reduction and clustering and (iii) non-parametric analysis of variance among types.

We selected variables related to the structural and functional features of farming systems as well as FHH socioeconomic characteristics in order to acquire a comprehensive view of the diversity of maize-based farming systems and livelihoods in the Western Highlands. In some cases, variables came directly from survey responses. In others, we constructed indices to integrate several answers. We selected 21 discriminant variables (19 continuous, 2 categorical) that statistically classified different rural livelihoods. Nineteen additional variables (18 numerical, 1 categorical) were used to further characterize the main features of the different farming systems. Categorical variables were ethnicity and household configuration (both discriminant), as well as traditional cropping systems management (for characterization). Annex 1 presents the 19 numerical variables used for the multivariate analysis (determinants), the 18 variables used for the description of farm types, and the three categorical variables.

Following Alvarez et al. (2014), we performed multivariate analysis to identify explanatory variables and to group farms into homogeneous types. Since our selected variables included categorical data, we used a factor analysis of mixed data (FAMD) (Pagès, 2004) and performed hierarchical clustering to form groups of relatively homogeneous FHH. We used the Scree Test (Cattell, 1966) to select the number of principal components to group FHH. Hierarchical Clustering on Principle Components (HCPC) builds a hierarchical tree and calculates one partition with an optimal number of clusters based on the partition with the highest relative loss of inertia.

A non-parametric analysis of variance identified the major differences in the discriminant and descriptive variables among clusters. As the numerical variables proved not to have a normal distribution (after Shapiro-Wilk normality test), we ran the Kruskal-Wallis rank sum test ANOVA and a *post hoc* test for detecting significant differences among pairs of clusters (Dunn's test). For categorical variables, we employed a chi-square test of independence for the contingency tables of farm types versus categorical variables.

Results

A Scree test indicated five dimensions from the FAMD analysis, which explained 35% of the variability. The main variables in each of the five dimensions are arable land available, travel time to urban areas, household size, household configuration and ethnicity.

The first two dimensions (Figure 6.2b) explain 18.2% of the variance. The right upper and lower quadrants differentiate the type-1 farms (Figure 6.2c circles) according to farm diversification, number of crops and sales intensity, as well as households headed by men from Ladino, Acateco, and Kanjobal ethnic groups. Type-2 farms are located in the most remote localities (lower right quadrant), and are mostly of Ixil, Jacalteco, and Popti ethnicity. Type-3 households are characterized by large family sizes and are located at high altitudes. These households are headed by men with a partner and are mainly Quiche and Mam ethnicity. The third dimension differentiates the other two types. Type 4 consists

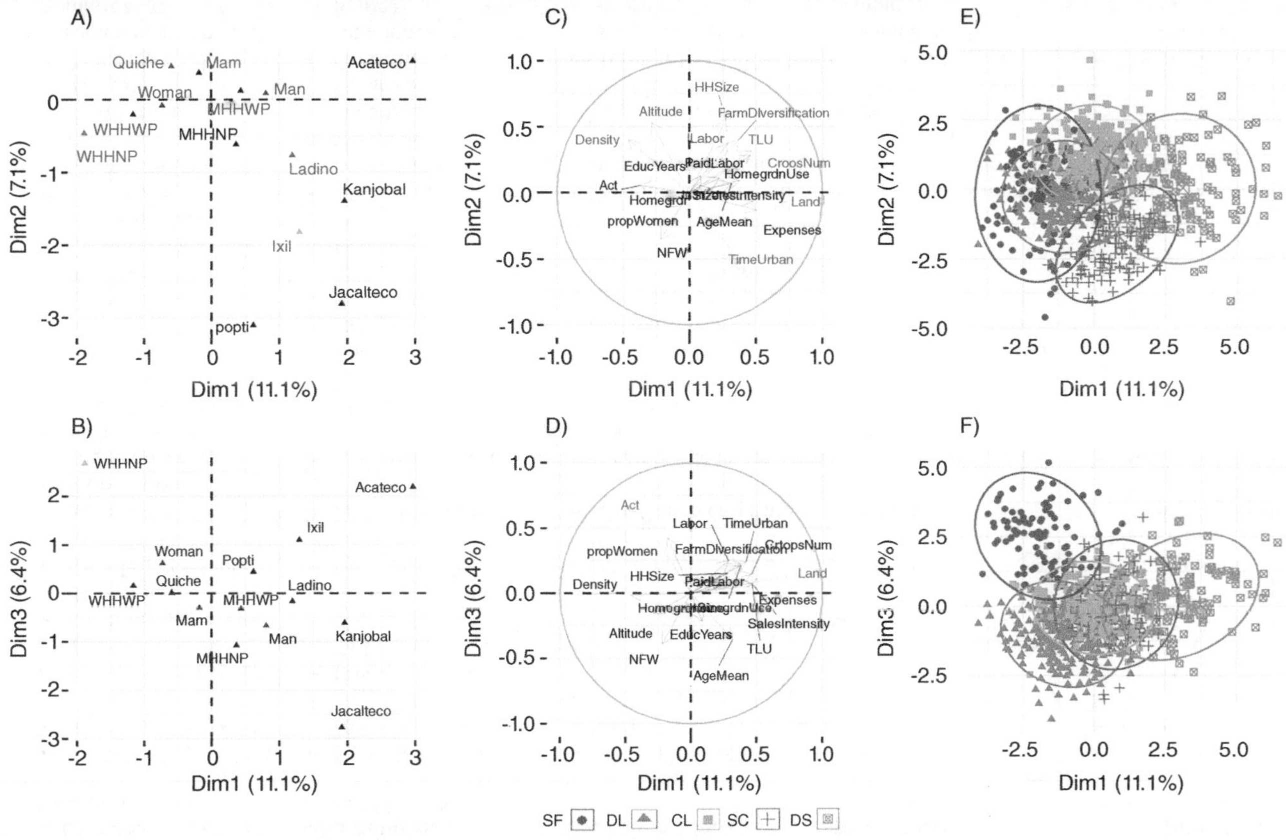

Figure 6.2 Factor analysis of mixed data (FAMD), biplots of quantitative (a,d) and qualitative (b,e) variables against the FAMD dimensions. The upper panels show the plane Dim 1 vs. Dim2; the lower panels show the plane Dim 1 vs. Dim 3. Panels c and d show the scatterplots of individuals against the dimensional spaces defined through FAMD.

Table 6.1 Types description by systems component; average type values of some variables are shown in parentheses

System Component	*Diversified livelihood (DL), infra-subsistence (N = 268, 32%)*	*Crop-Livestock (CL), infra-subsistence (N = 253, 30%)*	*Diversified subsistence (DS) (N = 123, 15%)*	*Specialized crop (SC), infra-subsistence (N = 114, 13%)*	*Self-consumption feminized (SF), infra-subsistence (N = 87, 10%)*
Structural	➢ Little land (0.15 Ha) ➢ Small families (4.5) ➢ Close to urban ➢ Few Livestock (0.22 TLU) ➢ Small home gardens (0.1 Ha)	➢ Densely populated (65 persons/ha) ➢ Large families (7.8) ➢ Livestock (0.44 TLU)	➢ Subsistence HH (0.7 Ha) ➢ Most livestock (0.91 TLU)	➢ Second in land availability (0.32) ➢ Least livestock (0.15 TLU) ➢ Furthest from urban ➢ Lowest altitude	➢ Little land (0.17 Ha) ➢ Large home garden ➢ Some livestock (0.22 TLU)
Functional	➢ Distribute HH labor equally between on/off farm ➢ Monocrop maize	➢ Most HH labor on farm ➢ Diversified cropping	➢ Most HH labor off farm ➢ Diversified cropping for sale (50%)	➢ On/off farm activities, but mostly for on farm ➢ Mostly monocrops	➢ Mostly work on farm ➢ Diversified cropping (Maize polycropping)
Social	➢ High education ➢ Male or female headed with spouse (typical of their ethnicity) ➢ Located in Quetzaltenango (Concepcion Chiquirichapa and San Juan Ostuncalco)	➢ Male headed with spouse ➢ Mostly Quiche ➢ Located in Totnicapan (Momostenango, Santa Maria Chiquimula, and San Bartolo)	➢ Male headed with spouse ➢ Acateco and Ladino ➢ Located in Huehuetenango (Chiantla, Concepcion Huista, and Todos Santos)	➢ Single male headed ➢ Ixil, Jacalteco, Ladino, and Popti ➢ Located in Quiche (Chaju, Nebaj, and San Juan Cotzal)	➢ Single woman headed ➢ Young and low education ➢ Located throughout
Maize production and use	➢ Low maize production, but high yield ➢ Feed little to livestock	➢ Modest management investment ➢ Good yields ➢ Feed little to livestock	➢ Spend most on maize management, but low yields ➢ More than others to feed livestock	➢ Much investment in maize management ➢ Medium yield ➢ Feed little to livestock	➢ Little investment in maize production ➢ Limited yield ➢ Feed little to livestock
Food security status	➢ 9 months food sufficiency (6.6 of maize)	➢ 9 months food sufficiency (6 of maize)	➢ 10 months food sufficiency (9 of maize)	➢ 10 months food sufficiency (7 of maize)	➢ 9 months food sufficiency (6 of maize)

of households with a high proportion of non-farm workers. Finally, type 5 (squares in Figure 6.2d) differs from the rest because households are headed by women.

Table 6.1 presents the main characteristics of each farm household type in relation to the structural, functional and social features, as well as some specific characteristics related to maize production and food security status. In Annex 1, the results of the non-parametric analysis of variance among groups are presented.

Different household types follow a geographical distribution. As Figure 6.3 illustrates, some types are concentrated in specific departments or municipalities. For instance, land-endowed types are found in the north. ***Diversified subsistence households (DS)*** are located in Huehuetenango. In a similar situation, ***Specialized Crop infra-subsistence households*** **(SC)** are mostly in North Quiche and include those most dedicated to agricultural activities for commercial and for self-consumption purposes. In contrast to the infra-subsistence farmers, ***crop-livestock infra-subsistence households (CL)***, ***diversified livelihood infra-subsistence households (DL)*** and ***self-consumption oriented infra-subsistence feminized households (SF)*** are located in the southern portion of the region, in Quetzaltenango Totonicapán, and south Quiche and in the north Quiche in the Cunen municipality. The DL type is best represented in Quetzaltenango but

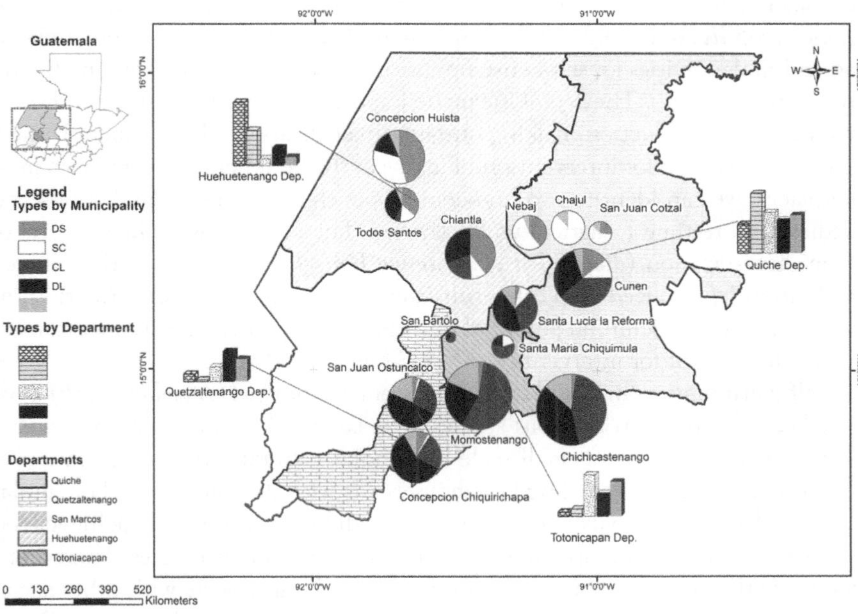

Figure 6.3 Farm types distribution in the departments (bar graphs) and municipalities (pie charts) surveyed in the Buena Milpa baseline. The size of each pie is relative to the number of surveys analyzed in each municipality. Farm types are: Diversified subsistence (DS), Specialized crop (SC) infra-subsistence, Crop-Livestock (CL) infra-subsistence, Diversified livelihood (DL), and Self-consumption feminized (SF) infra-subsistence.

is found in the smallest proportions in all the Departments, indicating that an important transition to non-farm activities as a livelihood strategy is emerging across the region. The SF type is well represented in all the departments and in most municipalities, except San Bartolo and Santa Maria Chiquimula. Thus, processes of agriculture feminization are combined with lowland access, low education levels and possibly migration of men, especially in the more populated areas in the South.

Discussion

Small-scale farmers from the Western Highlands Region have developed different livelihood strategies over the centuries. A key component of that development is the proximity of the region to urban settlements and the commercial and local production characteristics (Smith, 1989; van Etten, 2006). Based on that, these authors recognize three zones that explain maize production-consumption strategies of four of the farm typologies described in this chapter. In the Core Zone, ***diversified livelihood infra-subsistence households*** (DL) take advantage of the proximity to urban settlements to allocate their labor almost equally to on-/off-/non-farm activities. They cultivate maize in monocrop achieving high yields in very fertile but small plots. Interventions in this household type should consider labor trade-offs and land fragmentation challenges. In the Central Zone, ***crop-livestock infra-subsistence households*** (CL) invest more time in agricultural activities for self-consumption and for local markets than on off- and non-farm activities. They combine maize in polyculture with livestock and commercial crops production, making on-farm diversification a key element to take into consideration for interventions. Finally, in the Peripheral zone, far from marketplaces, we can identify two household types that have more land access and dedicate more time to agricultural activities. They follow two schemes, one of farm diversification (***diversified subsistence households***) and the other of specialization (***specialized crop infra-subsistence households***). This distinction (as well as the low agricultural potential that characterizes this zone) is important to take into account for interventions (MAGA et al., 2011).

Self-consumption–oriented infra-subsistence feminized households (SF) have not been described in previous studies. We assume that the addition of social variables to typologies of small-scale households facilitates the identification of this household type, revealing how households with different social but similar structural and functional characteristics adopt different strategies to achieve food security. Thus, gender is an important variable within the typologies. This is consistent with robust evidence indicating how gender intersects with other social identities to explain complex problems such as food insecurity (Sachs and Patel-Campillo, 2014). Our analysis reveals that gender intersects with household configuration in explaining household diversity. It also contributes to explaining the three processes (fragmentation, diversification and agriculture feminization) that define smallholders' heterogeneity, which are key for development interventions seeking to achieve food security.

Although land fragmentation occurs throughout the country, it has been more dramatic in the Western Highlands. Hellin et al. (2017) found that farmers from this region owning an average of 4.3 ha in the 1980s had only 0.4 ha in 2016. Land fragmentation represents a challenge for the maize production-consumption system of the region. Household types with the smallest land size (SF and DL) are food secure for fewer months of the year, even though their yields are more than 20% higher than *Diversified subsistence households* (DS), who have four times their land size. DS households are mainly headed by elder men with medium-level education and are majority *ladino* (non-indigenous). This indicates how social identities such as gender and ethnicity play important roles in this unequal situation. Women were excluded from land ownership, going back to Colonial times. Krishnamurthy and Schüssler (2007) contend that this gender inequality continues. Similarly, indigenous people were dispossessed of their land after the arrival of Spaniards in the 16th century, and they were used as free labor, a practice that continued during the 19th century (Villagrán, 2000). Today, women possess 6.5% of the land, even though they comprise 51.2% of the country's population (Krishnamurthy and Schüssler, 2007; Gobierno de Guatemala, 2013). Indigenous people represent 40% of the total population, but they own just 23.6% of the land (Krishnamurthy and Schüssler, 2007; Gobierno de Guatemala, 2013).

Some previous researchers have contended that introduction of non-traditional crops would promote food security (Immink and Alarcon, 1993; von Braun and Immink, 1994; Hamilton and Fischer, 2003). Others (Isakson, 2009; Webb et al., 2016) highlight the threats that crop diversification has on local food provision and consumption patterns. The complex relationship between small farmers and markets for their self-determination reflects what de Janvry (1975) calls "functional dualism."

Isakson (2009) concludes that "the resiliency of the Guatemalan peasantry is underpinned by its participation in multiple forms of economic provisioning" in which labor plays a key role. Our results confirm this assertion by demonstrating that all farm household types use different patterns to distribute their labor into on-, off- and non-farm activities. Moreover, household types reflect different levels of market integration as they produce commercial and self-consumption crops. Most household types follow a diversification scheme at a livelihood (DL), on-farm (CL and DS) or crop (SF) level. Of special interest are the *self-consumption–oriented infra-subsistence feminized households (SF)*, which are similar in land size and labor usage to the *diversified livelihood infra-subsistence households (DL)* but rely on crop, and not livelihood, diversification. Gender norms can explain this. Fraustro et al. (2013) find that rural Guatemalan women's access to paid work is constrained by gender norms related to roles and power asymmetries associated with women's autonomy.

Self-consumption–oriented infra-subsistence feminized households (SF) represent the feminization process of agriculture, "associated with an increase in the proportion of female households heads in rural areas, as well as male absence from the farm. This in turn is related to growing male migration or employment in off-farm pursuits" (Deere, 2005). These households tend to be headed by young women with the lowest education levels. They own the second-smallest land size and participate the least in non-farm activities. Moreover, their food security status is problematic. They are the ones who produce the lowest amount of maize and generally have the least number of months with maize and food availability. Their food provisioning strategy consists of on-farm diversification (especially maize in a polycropping system and home garden) for self-consumption and selling a little at markets. Although, in some cases, households receive remittances, women's need to increase their workload as they become responsible for both reproduction and production activities makes them more vulnerable to food insecurity. Because women typically have less access to land and do not engage in non-farm activities, they face constraints when trying to adapt to new challenges.

Our typology of maize farm households in the Western Highlands of Guatemala complements previous typological efforts by adopting an integral view of farming systems that adds a social dimension to the structural and functional components of the typology. It also contributes to understanding how social identities such as gender limit various food security strategies available to different households. Our findings indicate that processes related to farm diversification, land fragmentation and agriculture feminization appear important when seeking to explain households' heterogeneity in relation to their maize provision strategies. They also highlight the need to consider these strategies for developing agricultural development interventions that respond effectively to different food insecurity challenges.

Funding

The United States Agency for International Development (USAID) through its Global Hunger and Food Security Initiative, Feed the Future, supported the Project Feed the Future Buena Milpa.

Literature cited

Alvarez, S., Paas, W., Descheemaeker, K., Tittonell, P.A. and Groot, J. (2014). *Constructing typologies, a way to deal with farm diversity: General guidelines for the humidtropics*. Report for the CGIAR Research Program on Integrated Systems for the Humid Tropics. Plant Sciences Group, Wageningen University, the Netherlands. Humidtropics, a CGIAR Research Program led by IITA.

Baral, S. and Bardhan, D. (2016). Multivariate typology of milk producing households in Uttarakhand hills: Explaining profitability in dairy farming. *Indian Journal of Agricultural Economics*, 71, pp. 160–175.

Bellow, J.G., Hudson, R.F. and Nair, P.K.R. (2008). Adoption potential of fruit-tree-based agroforestry on small farms in the subtropical highlands. *Agroforestry Systems*, 73, pp. 23–36. doi:https://doi.org/10.1007/s10457-008-9105-x.

Berre, D., Baidron, F., Kassie, M., Craufurd, P. and Lopez-Ridaura, S. (2016). Different ways to cut a cake: Comparing expert-based and statistical typologies to target sustainable intensification technologies, a case-study in Southern Ethiopia. *Experimental Agriculture*, pp. 1–17. doi:https://doi.org/, doi:10.1017/S0014479716000727.

Bidogeza, J.C., Berentsen, P.B.M., De Graaff, J. and Oude Lansink, A.G.J.M. (2009). A typology of farm households for the Umutara Province in Rwanda. *Food Security*, 1, pp. 321–335. doi:https://doi.org/10.1007/s12571-009-0029-8.

Bruni, L., Fuentes, A. and Rosada, T. (2009). *Dynamics of inequality in Guatemala, UNDP project "Markets, the state, and the dynamics of inequality: How to advance inclusive growth," coordinated by Luis Felipe López-Calva and Nora Lustig*. Available at: http://undp.economiccluster-lac. org/.

Cattell, R.B. (1966). The screen test for the number of factors. *Multivariate Behavioral Research*, 1, pp. 245–276.

Cochet, H. (2015). *How to build farm typologies? BT – comparative agriculture*. H. Cochet, ed. Dordrecht, Netherlands: Springer, pp. 109–114. doi:https://doi.org/10.1007/978-94-017-9828-0_9.

Deere, C.D. (2005). *The feminization of agriculture? Economic restructuring in Rural Latin America*. Economic restructuring in rural Latin America (No. 1). UNRISD Occasional Paper. UNRISD, Palais des Nations, 1211 Geneva 10, Switzerland. www.unrisd.org/publications/opgp1.

de Janvry, A. (1975). The political economy of rural development in Latin America: An interpretation. *American Journal of Agricultural Economics*, 57, pp. 490–499. doi:https://doi.org/10.2307/1238412.

de Janvry, A. and Sadoulet, E. (2010). The global food crisis and Guatemala: What crisis and for whom? *World Development*, 38, pp. 1328–1339. doi:https://doi.org/10.1016/j.worlddev.2010.02.008.

Elliott, E.D. (1989). *A history of land tenure in the Ixil triangle*. La Antigua, Guatemala: CIRMA.

Fraustro, M., Peláez-Ponce, A.V. and Zapil, S. (2013). *El nuevo perfil de las mujeres rurales jóvenes en Guatemala*. Lima: IEP; Nuevas Trenzas. (Documento de Trabajo, 180. Serie Programa Nuevas Trenzas, 5).

Fuentes-López, M.R., Van Etten, J., Vivero Pol, J.L. and Ortega Aparicio, A. (2005). *Maíz para Guatemala: Propuesta para la Reactivación de la Cadena Agroalimentaria del Maíz Blanco y Amarillo, SERIE 'PESA Investigación,' n 1*. Guatemala City: FAO.

Gálvez, J. (2000). *Analisis ambiental general del Altiplano Occidental y del proyecto MIRNA*. Guatemala City: MAGA, CONAP, INAB.

Gobierno de Guatemala. (2013). *Caracterizacion Estadistica República de Guatemala*. Instituto Nacional de Estadística, Gobierno de Guatemala, República de Guatemala.

Gobierno de Guatemala. (2012a). *Caracterizacion Estadistica República de Guatemala*. Instituto Nacional de Estadística, Gobierno de Guatemala, República de Guatemala.

Gobierno de Guatemala. (2012b). *Programa de Agricultura Familiar para el Fortalecimiento de la Economía Campesina (PAFFEC 2012–2015), elaborado por el Ministerio de Agricultura, Ganadería y Alimentación, MAGA, con el apoyo de la Organización de las Naciones Unidas para la Agricultura y la Alimentación*. FAO.

Gobierno de Guatemala. (2005). *Política Nacional de Seguridad Alimentaria y Nutricional*. SEGEPLAN, Gobierno de Guatemala. Guatemala.

Guardiola, J., Cano, V.G. and Pol, J.L.V. (2006). *La seguridad alimentaria: estimación de índices de vulnerabilidad en Guatemala*. VIII Reunión de Economía Mundial. 20–22 Abril. Alicante: Universidad de Alicante.

Hamilton, S. and Fischer, E.F. (2003). Non-traditional exports in Highlands Guatemala: Understandings of risk and perceptions of change. *Latin American Research Review*, 38(3), pp. 82–110. University of Texas Press, Austin, Tx.

Hellin, J., Cox, R. and López-Ridaura, S. (2017). Maize diversity, market access, and poverty reduction in the Western highlands of Guatemala Mountain. *Research and Development*, 37, pp. 188–197.

Immink, M.D.C. and Alarcon, J.A. (1993). Household income, food availability, and commercial crop production by smallholder farmers in the Western Highlands of Guatemala. *Economic Development and Cultural Change*, 41, pp. 319–342. doi: https://doi.org/10.1086/452013.

Isakson, S.R. (2014). Maize diversity and the political economy of agrarian restructuring in Guatemala. *Journal of Agrarian Change*, 14(3), pp. 347–379. doi:https://doi.org/10.1111/joac.12023.

Isakson, S.R. (2009). No hay ganancia en la milpa: The agrarian question, food sovereignty, and the on-farm conservation of agrobiodiversity in the Guatemalan highlands. *Journal of Peasant Studies*, 36, pp. 725–759. doi:https://doi.org/10.1080/03066150903353876.

Krishnamurthy, A. and Schüssler, R. (2007). *Mujeres toman el poder de la tierra: acceso a la tierra como una estrategia de empoderamiento de mujeres indígenas en Guatemala*. Heidelberg, Alemania: FIAN Internacional.

MAGA, SESAN, APSAN, AID. (2011). *Diagnóstico de la región de occidente de Guatemala*. MAGA, Gobierno de Guatemala, Guatemala City.

Mc Creery, D. (1989). Tierra, mano de obra y violencia en el Altiplano Guatemalteco: San Juan Ixcoy. *Revistade Historia*, 19, pp. 19–35, Costa. Rica.

Meinzen-Dick, R., Behrman, J., Menon, P. and Quisumbing, A. (2012). *Gender: A key dimension linking agricultural programs to improved nutrition and health in: Reshaping agriculture fornutrition and healt*. S. Fan and R. Pandya-Lorch. International Food Policy Research Institute, pp. 135–144.

Moran-Taylor, M.J. and Taylor, M.J. (2010). Land and leña: Linking transnational migration, natural resources, and the environment in Guatemala. *Population and Environment*, 32, pp. 198–215. doi:https://doi.org/10.1007/s11111-010-0125-x.

Pagès, J. (2004). Analyse factorielle de données mixtes. *Revuede Statistique. appliquée*, 52(4), pp. 93–111.

Sachs, C. and Patel-Campillo, A. (2014). Feminist food justice: Crafting a new vision. *Feminist Studies*, 40(2), pp. 396–410.

Samberg, L.H., Gerber, J.S., Ramankutty, N., Herrero, M. and West, P.C. (2016). Subnational distribution of average farm size and smallholder contributions to global food production subnational distribution of average farm size and smallholder contributions to global food production. *Environmental Research Letters*, 11(12), pp. 1–11. doi:https://doi.org/10.1088/1748-9326/11/12/124010.

Sigüenza Ramírez, P., Winkler, K., Monzón, R., Gauster, S., Dürr, J. and Ozaeta, J.P. (2010). *Nuestro maíz, nuestro futuro: Estudios para la reactivación de la producción nacional de maíz en Guatemala*. Instituto de Estudios Agrarios y Rurales, IDEAR, Coordinación de ONG y Cooperativas,CONGCOOP. Magna Terra Editores, Guatemala City.

Smith, C.A. (1989). Survival strategies among petty commodity producers in Guatemala. *International Labour Review*, 128(6), p. 791.

Steinberg, M.K. and Taylor, M.J. (2009). The direct and indirect impacts of population growth and economic development on maize (Zea mays L.) diversity in highland Guatemala. *Area*, 41(1), pp. 72–81. doi:https://doi.org/10.1111/j.1475-4762.2008.00850.x.

Taylor, J.E., Yuñez-Naude, A., Jesurun-Clements, N., Huard, A., Sanchez, M.A., Alvarez, V.M. and Baumesiter, E. (2006). *Los posibles efectos de la liberalización comercial en los hogares rurales Centroamericanos a partir de un modelo desagregado para la economía rural*. Banco Interamericano de Desarrollo.

Tittonell, P., Muriuki, A., Shepherd, K.D., Mugendi, D., Kaizzi, K.C., Okeyo, J., Verchot, L., Coe, R. and Vanlauwe, B. (2010). The diversity of rural livelihoods and their influence on soil fertility in agricultural systems of East Africa -A typology of smallholder farms. *Agricultural Systems*, 103(2), pp. 83–97. doi:https://doi.org/10.1016/j.agsy.2009.10.001.

Tittonell, P., Vanlauwe, B., Leffelaar, P.A., Shepherd, K.D. and Giller, K.E. (2005). Exploring diversity in soil fertility management of smallholder farms in western Kenya: II. Within-farm variability in resource allocation, nutrient flows and soil fertility status. *Agriculture Ecosystems & Environment*, 110(3–4), pp. 166–184. doi:https://doi.org/10.1016/j.agee.2005.04.003.

USAID/Guatemala. (2009). *Evaluación rápida del sector agrícola guatemalteco y su estado para abordar los retos de seguridad alimentaria del país: Informe Final Bajo el Programa de Comercio y Competitividad. USAID*. Ciudad de Guatemala, Guatemala.

van Etten, J. (2006). Molding maize: The shaping of a crop diversity landscape in the Western Highlands of Guatemala. *Journal of Historical Geography*, 32(4), pp. 689–711. doi: https://doi.org/10.1016/j.jhg.2005.12.002.

van Etten, J. and Fuentes, M.R. (2004). La crisis del maíz en Guatemala: las importaciones de maíz y la agricultura familiar. *Anuariode Estudios Centroamericanos*, 30(1), pp. 51–66.

Villagrán, K. (2000). Guatemala, la fuerza incluyente del desarrollo humano: La fuerza incluyente del desarrollo humano: informe de desarrollo humano 2000. Sistema de Naciones Unidas en Guatemala.Ciudad de Guatemala.

von Braun, J. and Immink, M.D.C. (1994). Nontraditional vegetable crops and food security among smallholder farmers in Guatemala. In: J. von Braun and E.T. Kennedy, eds., *Agricultural commercialization, economic development, and nutrition*, Chapter 12, pp. 189–203. Baltimore, MD: Published for the International Food Policy Research Institute (IFPRI) by Johns Hopkins University Press, Baltimore MD. http://ebrary.ifpri.org/cdm/ref/collection/p15738coll2/id/129377

Webb, M.F., Chary, A.N., De Vries, T.T., Davis, S., Dykstra, M., Flood, D., Rhodes, M.H. and Rohloff, P. (2016). Exploring mechanisms of food insecurity in indigenous agricultural communities in Guatemala: A mixed methods study. *BMC Nutrition*, 2(1), p. 55. doi: https://doi.org/10.1186/s40795-016-0091-5.

Wilken, G.C. (1971). Food-producing systems available to the ancient maya. *American Antiquity*, 36(4), pp. 432–448. doi: https://doi.org/DOI: 10.2307/278462.

Annex I

Statistical comparison of farm HH types

Table 6.2 Average values of structural, functional and social variables that characterize farm types and total sample

	Variable	*Variable Acronym*	*Units*	*DL n = 268*		*CL n = 253*		*DS n = 123*		*SC n = 114*		*SF n = 87*		*Descriptive/ Discriminant variable*	*Total sample Mean*	*Sd.*
Structural	Arable land	Land	Ha	0.15	d	0.21	c	0.7	a	0.32	b	0.17	d	Discriminant	0.27	0.28
	Density	Density	Persons/Ha	49	b	65	a	13	d	23	c	54	b	Discriminant	46	51
	Household family size	HHSize	Persons	4.5	c	7.8	a	6.1	b	5.8	b	4.7	c	Discriminant	6	2
	Home garden size	HomegrdnSize	Ha	0.01	c	0.01	b	0.02	a	0.01	bc	0.01	bc	Discriminant	0.01	0.04
	Tropical livestock units	TLU	TLU	0.22	c	0.44	b	0.91	a	0.15	c	0.22	c	Discriminant	0.37	0.58
	Altitude	Altitude	Meters	2385	a	2319	b	2212	b	1892	c	2346	ab	Discriminant	2269	348
	Time from village to urban areas	TimeUrban	Minutes	27	c	28.04	bc	65	a	76	a	37	b	Discriminant	41	32
	Number of crops grown	CropsNum	Crops	1.8	c	2.1	b	3.2	a	1.7	c	2	bc	Discriminant	2	1
	*Cropping system[F]	CropSys	7 crop systems	MM: 4		MP: 2		CP: 3.8, OC: 4.9, PM:2.7, PP: 2.2		CM: 3.3, MM: 2.7		MP: 0.5		Descriptive	MP:63.3, MM:29.8, CM:0.4, CP: 0.36, OC:2.25, PM:1.66, PP: 0.12	
	Use of home garden	HomegrdnUse	Num.of activities	1.2	c	2	b	2.9	a	1.5	c	1.6	bc	Discriminant	1.8	1.6
	Intensity of farm produce sales	SalesIntensity	%	6	bc	4	bc	50	a	8	b	0.3	c	Discriminant	11	29
	Farm activities diversification	Farm diversification	Index:1–4	2.4	d	3.2	b	3.6	a	2.2	d	2.8	c	Discriminant	2.81	1.01
	Money spent on agronomic activities	Expenses	Quetzales	212	b	271	b	675	a	588	a	261	b	Discriminant	353	369
	Activity of women in agricultural activities	Act	Five points scale	0.29	c	0.32	b	0.24	c	0.23	c	0.63	a	Discriminant	0.32	0.2

Functional	Paid labor	PaidLabor	Persons	1.4	b	2.5	ab	3.8	a	2	b	2.4	ab	Discriminant	2	4
	On-farm labor force proportion	ONLFP	persons/ HHtotal	0.66	a	0.48	d	0.6	ab	0.52	cd	0.57	bc	Descriptive	0.57	0.24
	Off-farm labor force proportion	OFFLFP	persons/ HHtotal	0.59	a	0.35	c	0.45	b	0.43	b	0.42	bc	Descriptive	0.46	0.23
	Non-farm labor force proportion	NFLFP	persons/ HHtotal	0.62	a	0.34	c	0.41	b	0.44	b	0.4	bc	Discriminant	0.46	0.23
	Percent of land under maize production	POLUMAIZE	%	95	a	96	a	81	b	93	a	99	a	Descriptive	94	18
	Production of Maize in Kg	MaizeKg	Kg	273	d	319	c	768	a	481	b	242	d	Descriptive	384	389
	Yield of Maize	MaizeYield	Kg/Ha	2.27	a	2.03	ab	1.73	b	1.88	ab	2.16	ab	Descriptive	2.04	1.65
	Proportion of maize consumed by family	MaizeConsProp	%	87	a	82	b	76	c	90	a	87	ab	Descriptive	85	17
	Proportion of maize consumed by livestock	MaizeLvstk ConsProp	%	13	c	19	ab	22	a	13	c	18	bc	Descriptive	16	14
	Proportion of maize sold	MaizeSoldProp	%	1.5	b	0.3	b	6	a	1	b	0	b	Descriptive	1.5	8
	Production of Bean in Kg	BeanKg	Kg	25	c	39	b	63	a	37	bc	8	c	Descriptive	35	119
	Yield of Bean	BeanYield	Kg/Ha	0.21	a	0.21	a	0.17	b	0.16	b	0.09	c	Descriptive	0.19	0.58
	Proportion of bean consumed	BeanConsProp	%	40	c	60	ab	71	a	43	c	52	bc	Descriptive	52	49
	Proportion of bean sold	BeanSoldProp	%	0.5	b	1	b	6	a	2	ab	0	b	Descriptive	1.6	9.3
	Daily average meals	DailyAverage Meals	Meals	2.98	ab	2.99	ab	3	a	2.9	b	3.03	ab	Descriptive	3	0.4
	Number of food aid months	Months FoodAid	Months	0.34	b	0.56	ab	1.2	a	0.75	ab	0.16	ab	Descriptive	0.6	2.3
	Months of food sufficiency	Months FoodSuff	Months	9.4	a	8.9	a	10	a	9.6	a	9.2	a	Descriptive	9.4	3.3
	Months of maize self sufficiency	MaizeSelfSuff	Months	6.6	bc	5.9	c	8.6	a	7.3	ab	5.9	c	Descriptive	6.7	3.7
	Months of bean self sufficiency	BeanSelfSuff	Months	1.9	c	2.9	b	5.3	a	2.9	bc	1.6	c	Descriptive	2.9	3.7

(Continued)

Table 6.2 (Continued)

	Variable	*Variable Acronym*	*Units*	*DL n = 268*		*CL n = 253*		*DS n = 123*		*SC n = 114*		*SF n = 87*		*Descriptive/ Discriminant variable*	*Total sample Mean*	*Sd.*
Social	Mean age of couple	AgeMean	Years	36	b	38	ab	42	a	34	b	24	c	Discriminant	36	14
	Mean scholar level of couple	EducYears	Years	7.1	a	6.3	a	6.8	a	5.4	ab	3.3	b	Discriminant	6	5
	Proportion of women in HH	propWomen	%	0.51	bc	0.48	c	0.52	b	0.47	C	0.67	a	Discriminant	0.51	0.17
	+Household Configuration[F]	HHConfig	4 categories	MHHWP:3		MHHWP:2		MHHWP:3		MHHNP: 5		WHHNP: 23		Discriminant	%= MHHWP:74, WHHWP:12.8, WHHNP: 9.2,MHHNP:3.1	
	Ethnicity[F]	Ethnicity	7 groups	Mam: 4		Quiche: 7		Acateco :3.8, Ladino :6.4		Ixil: 13.8, Jacal-teco: 1.4, Ladino 3.2, Popti: 3.3		None		Discriminant	%= Quiche: 48.4, Mam:25.2, Ladino:15.4, Ixil:9.3, Kanjobal:0.6, Popti:0.5, Acateco:0.4, Jacalteco:0.2	

Lowercase letters denote different groups at $p < 0.05$ for numerical variables. For factors, the significant relationships between types and factor levels detected by chi-square tests are indicated, along with the Pearson's residuals value.
*MM = Maize Monocrop, MP = Maize Polyculture, CP = Coffee PolyCulture, OC = Other Crops, PM = Potato Monoculture, PP = Potato Polyculture, CM = Coffee Monocrop.
+MHHWP = Men headed households with a partner, WHHWP =women headed households with a partner, WHHNP =women headed households with no partner, MHHNP = men headed households with no partner. [F]Factor variables the rest are numeric.

7 "A Bird Locked in a Cage"

Hmong young women's lives after marriage in northern Vietnam

Nozomi Kawarazuka, Nguyen Thi Van Anh, Vu Xuan Thai and Pham Huu Thuong

Introduction

Ethnic minorities live in the mountain areas across five Southeast Asian countries and China. They used to engage in shifting cultivation and have very different cultural and agricultural histories from the rice-based lowland populations in this region (Scott, 2009). In Vietnam, poverty is concentrated in those mountain areas. The state government and international organizations have been providing specific agricultural support for these minorities through introducing new technologies such as hybrid rice and maize seeds and new cash commodities like coffee, rubber and cassava to increase household incomes. It is, however, recognized that the underlying problems of poverty for the minorities are not only economic constraints but more importantly cultural and social norms that foster social exclusion (World Bank, 2009). The literature of political ecology shows rather negative impacts of agrarian transitions on ethnic minorities, creating further gaps between the majority and minority (e.g., Minot and Baulch, 2005; Doutriaux et al., 2008; Friederichsen and Neef, 2010). In this argument, however, less attention has been paid to the impacts of ongoing changes in agricultural technology on gender relations, in particular, for the younger generation. Ethnic-minority women farmers are active agents: they partly adapt to modern agriculture by negotiating with men, seek other livelihood opportunities, and still sustain subsistence farming for household food security. In this process of adaptation to change, gender roles and relationships are contested and reconfigured (Lindeborg, 2012; Turner, 2012; Bonnin and Turner, 2014). Recently, the living conditions in the mountain areas in Vietnam have been rapidly changing due to improved infrastructure and access to information, including the internet; improved education levels; and ongoing rural–urban migration, all of which reduce social and cultural barriers. Young ethnic minority men and women in Vietnam thus live in a time of transition.

If an end goal of agricultural development for the minorities in Vietnam is not simply increasing average incomes and production but improving the quality of the lives of the poor, half of whom are women, we need to closely explore how gender relations are changing and how these changes affect in particular young ethnic minorities and their perceptions of current agriculture. This is an entry point for providing more inclusive and equitable development support.

The present study seeks to explore young women's and men's experiences in the two related domains of marriage and agriculture, with their multiple identities and desires, in the context where gender norms, gender relationships and livelihood strategies are changing.

We address three research questions:

1 How do young women negotiate over prevailing social norms and values in marriage and family relations?
2 How do young women's intersectional position shape the ways they engage with agriculture?
3 How much do young women have access to labor-saving technology and take advantage of those opportunities for agricultural and domestic work?

The following section explains a conceptual framework drawn from social theories of gender relations and the feminist theory of technology to understand the complex situation of male and female youth in the ethnic-minority context. We then explain participatory research methods and briefly describe research context. Key findings are divided into three subsections: (1) Hmong young women's gender relations in pre- and post-marriage processes; (2) Hmong young women's and men's negotiations with their families over agricultural labor and their involvement in non-agricultural business; and (3) gendered access to and perceptions on labor-saving technologies. While the study conducted similar numbers of interviews for men, it is rather focused on women's stories and their voices. This is because young women appear to be greatly influenced by ongoing technology change in agriculture, while their voices have not been sufficiently heard in agricultural research.

Conceptual framework

Earlier studies with a feminist epistemological approach recognized gender as power constructed by everyday practices and discourse practices (e.g., West and Zimmerman, 1987; Butler, 1990). Gendered everyday practices create symbolized meanings and gendered emotions. Gendered practices include productive, reproductive and social activities and discourse practices (gendered words used in everyday conversations and in media), shaping power relations between men and women as well as among men and among women (Butler, 1990; Connell, 2009). Gendered identities as to what women or men should be and should do are also influenced by gendered emotions of pleasure and guilt, which makes women's agency very complex (Whitehead, 1981; Kandiyoti, 1988; Kabeer, 2000). In the patriarchal society, women do not explicitly challenge power, as it is contrary to how women should behave and also, they know what approaches best work in their subordinated situations on which they depend (Schutte, 2014). The literature on gender relations and women's agency in agriculture and environmental conservation shows that women are often active agents despite their subordination, and their preferences, choices and incentives in agriculture and environmental conservation are closely associated with their gendered positions and family relationships and how they manage them (Jackson, 1998; Nightingale, 2006; Resurreccion, 2006; Lindeborg, 2012; Bonnin and Turner, 2014; Elmhirst et al., 2017). By critically looking at gender relations and young women's agency, this study contributes to the literature that challenges the concept of the "household" as a minimum unit for analyzing impacts of agricultural change and/or agricultural interventions (e.g., Guyer and Peters, 1987; Hart, 1997; Carr, 2005).

The second part of this study focuses on the issue of women's labor in patriarchal contexts and explores agricultural mechanization through critical gender analysis. Agricultural mechanization is defined in this study as a process of adopting labor-saving machines and tools for agriculture. The literature on feminist theories of technology (e.g., Faulkner, 2001; Wajcman, 2010) explores the process of technology development and its social consequences. Wajcman (2010) suggests that the relationships between technology and gender are two ways: technology is influenced by gender relations and technology influences gender relations. The general example of the former in agriculture is that technology innovations are often proposed by male researchers in response to priority constraints identified by male farmers: in consequence, the patriarchal society produces and reproduces male-oriented technology. An example of the latter (technology influences gender relations) is that the introduction of agricultural machines such as tractors and transplanting machines have resulted in men gaining power and prestige both materially and symbolically. The patriarchal society can thereby be not only sustained but also empowered. Understanding the social construction of technology including the gender dimension allows us to explore agricultural technology in more critical ways with the existing power relations not only through gender but also age and ethnicity, thus using an intersectional approach. In this study, particular attention is paid to understand specific issues of agricultural technology faced by ethnic-minority young women and men in a mountainous remote area.

Research methods and contexts

Research methods

A two-week fieldwork study was conducted in a Hmong village in Dien Bien Province in May 2018. Five researchers (two men and three women) were involved in the fieldwork. Prior to this field study, a rapid gender analysis was conducted in the same village, in which three researchers from this research team were involved (ICRAF-the Gender SRA, 2018). This was the second time to visit the village, and data were collected through: participatory videos (three young men and three young women farmers); a video screening and discussion event (14 men and 14 women); a photo-taking activity (four men and four women), four focus group discussions (better-off and poor, men and women each); and in-depth interviews (eight men and six women). The participants were from various socioeconomic and family conditions and were selected based on advice from the village leader. Their ages ranged from 16 to 29, and they were all married, reflecting the village context. The male and female participants were intentionally selected from different households to make the interviewees comfortable to speak about their conjugal relationships.

Before conducting in-depth interviews and focus group discussions (FGDs), two short films were made by a group of young men and women farmers, respectively, as an icebreaking activity. The Hmong women in the study site are not used to speaking in public, and many of them are very shy with outsiders, and this activity helped them to feel more confident in communication. We therefore needed to have this activity at the beginning of the fieldwork. It also helped us to explore young men's and women's general perceptions of their lives and livelihoods. The women filmed scenes with their children around, cooking in the kitchen and standing in front of water facilities, while the men focused more on information technologies, agriculture and infrastructure, all reflecting their highly gendered everyday roles and responsibilities. The research team organized a video screening event for the villagers and some district and commune officers, who all enjoyed watching them, and based on these two films and reactions to them, we modified the questionnaires for focus group discussions (FGDs) and in-depth interviews. A photo activity was conducted to explore men and women's access to labor-saving technology. The FGDs explored gender norms, ethnic identities, the use of labor-saving technology, and the social lives of men and women, while the in-depth interviews focused on biographies and personal narratives (Suarez-Ortega, 2013). Narratives covered family relations, social pressures, perceived values of education, challenges and interests in agriculture and desires for their daughters and sons.

Prior to all the data collection activities, verbal informed consent was obtained from all participants, during which confidentiality (e.g., not disclosing personal data to other villagers and not using their names in the reports) was emphasized. The FGDs and in-depth interviews were recorded with the agreement of all participants, with all recorded information then transcribed in Vietnamese and translated into English. The research team then met to discuss the key findings. Using thematic content analysis (Smith, 1992; Braun and Clarke, 2006), the first author analyzed the data collected during the fieldwork and led the writing of this article. All names of individuals presented in this chapter have been changed to protect anonymity.

Research context

This study site is located around 20 km up from the district town in Dien Bien province, Vietnam. The village is dominated by the White Hmong ethnic group, a subgroup of the Hmong whose ethnic dress is white. There are some women who are originally from red and black Hmong ethnic groups.

In the study village, both men and women marry in their late teens, which overlaps with their education period. Most girls stop studying after marriage, and the few who continue to study up to high school or college after marriage do so with support from their husbands, in-laws and birth families. Men tend to complete at least 9th grade or higher (12th) regardless of marital status. There is a patrilocal tradition in the community, whereby young couples usually live with the husband's parents and economically depend on them. Daughters-in-law are often expected to work hard as family laborers in both domestic and farming work.

Hmong people used to engage in subsistence farming of upland rice and maize. As part of the government policy to support ethnic minorities in the mountain areas, cash crops, such as hybrid maize, cassava, coffee and fruits were introduced over a decade ago in this region, and the current main agricultural income sources are coffee and local fruits. Annual household income ranges from 20–100 million Vietnamese dong ($850 to $4,500). Most of their agricultural activities depend on family labor, and only a few better-off households own agricultural machines such as cultivators, threshing machines and mills. Many young men work outside the village seasonally to earn additional income as laborers in the construction or industrial sectors or working for better-off farmers in neighboring villages.

Motorcycles are the major source of transport, and both men and women use them. Most households have normal mobile phones. Smartphones are increasingly used for better-off households, which men have more access to than women from the same household. The Internet is available in some parts of village areas, and around 70% of men and 20% of women in the late teens and early twenties have Facebook accounts. Although women do most domestic work, men sometimes help, and it is not uncommon to see men carrying babies on their backs at home, for example.

Key findings

Hmong women's experiences of marriage and their lives after marriage

Unlike their parents' generation, the Hmong village teenagers in the study site enjoy their social lives, checking Facebook, listening to V-pop on phones, Hmong music imported from Thailand and Laos, watching Korean comedy movies, wearing modern dresses in addition to ethnic costume and protecting their skins from the sun. Young women sometimes also use the internet to look for Hmong recipes and browse shopping sites, although not to actually buy anything. Social networking sites allow them to explore the variety of women's lives and livelihoods in the world, and so they are aware that there are Vietnamese women who conduct and even own businesses. Some Hmong girls admire such economically independent women.

However, their period as "youth" does not last long. In the study site, Hmong girls marry in their mid- or late teens, although their marriage certificate will be made at 18 years or later, as marriageable age is 18 and 20 for women and men respectively in Vietnam. They often accept a marriage proposal after a very short time. Ms Hang recalled her marriage at 15. Her husband worked at the same cement factory as her brother, who gave him her phone number, which is how their relationship began. In his first call, he introduced himself as a single man, a friend of her brother. She introduced herself, giving him her name and address. After a few more phone conversations, he visited her and proposed. Upon her immediate acceptance, he asked her mother for consent, and on the same day, Ms Hang went to his house and stayed there for three days. It is a custom in the village for a bride to stay in her prospective groom's house for a few days before an official wedding ceremony. When we asked her whether she had the option of leaving him before the end of her three-day stay, she said, "No, if I'd left him, people in my village would've seen me as a divorcee and I would then have lost value as a woman." In the study site, divorce is taboo for women regardless of age, who would be socially stigmatized. Men have no such problem, and they can easily remarry. She said that she had already firmly determined to marry him before visiting his house, particularly as she did not go to school and her parents did not allow her to go out much, so she had few opportunities to meet young men. When the chance of marriage came up, she saw it as fate.

Similarly, Ms Anh met her husband when she was in the ninth grade at school in the district town. It was during the New Year holiday in February, and she watched him playing football. After the game, he kept looking at her. He eventually found her on Facebook and sent her a friend request. She accepted, and from then they texted each other on their phones. Later that year, he visited her house and asked her to visit him to help pick mountain apples in his parents' farm. When she arrived at his house, he proposed to her. She did not know what she should say but stayed with him overnight, thereby accepting his proposal. The following morning, she went back to her house and told her mother that she was going to marry. Her mother was unaware that her daughter had not been home the previous night but accepted her decision. She quit her study and married in that year. Like Ms Anh, many Hmong women stop studying when they marry, even at around 15 to 17.

The custom of women's teenage marriage is partly rooted in the family pressure. Mr Tan married when he was 18 years old. He admitted that he was still immature at that time and depended on his parents both financially and mentally, "but I needed to marry to help my parents, who were getting old. Our family needed to have a wife who could help with the farming in the field." As is clear, a wife is often seen as an extra family laborer, and physically strong, hardworking women are often preferred. In such a situation, married life is often very hard for the teenage girls. According to the focus group discussions with both men and women, daughters-in-law are expected to sacrifice their time and labor to their husband, children and in-laws and be humble and obey their husbands and in-laws without argument. Ms Thu explained, "When a Hmong woman marries, her in-law family's needs should be paramount, above her own or her birth family. She always has to decide things based on what her in-laws want."

Ms Phuong said, "Being married is like being a bird locked in a cage. There is a lot of work to do and no time (to go out)." Similarly, Ms Anh recalled, "I did not feel that I was at home. When I first came and lived here, I did not know anyone and had no one to chat with. I was too shy to ask anyone how to manage tasks, and there were endless tasks. When I solved one issue, the next challenge popped up." Ms Hang said, "I worried whether my parents-in-law liked me. I hadn't gone to school and I did not know how to cook to their liking. I hesitated to ask my mother-in-law so I had to ask my husband everything. One day, my husband told me that his parents had wanted a daughter-in-law who was not educated because they wanted a person who could work on the farm, so I was reassured." Ms Anh said, "I have never gone on a trip with my husband. Even when we have the time, he plays soccer, and games in an Internet cafe. He told me that whenever he is upset, gaming can help him to get over it. But when I am depressed or crying, he doesn't console me. I sometimes can't stop weeping alone. We don't quarrel. He won't listen to me. I let him go out to play video games as long as he wants, usually a few hours."

Nevertheless, women usually manage and find support from others. Ms Anh said, "Eventually I found a supportive woman whom I called my aunt and I was able to ask her when I did not know how to do something. Now, there is nothing in this household that I am unsure of how to do, and I am not afraid of asking my in-laws." Ms Quynh said, "My mother-in-law is so selfish she doesn't want to give even one cent for her grandchildren. My father-in-law is always drunk and shouts at me. I did not know my husband's parents are like that. They don't care even when I'm sick. However, I have my parents who support me and my children in times of needs."

According to the in-depth interviews and focus group discussions, in the typical household in the study site, a wife's parents in-law are major decision makers, although the degree of decision-making power varies. Many men perceive that their wives have less knowledge and capacity to make household decisions than themselves, while women consider that they make decisions about many things in many ways. This gap seems to come from the women's own interpretation of decision making in which they find room for negotiation, while men talk about how men are expected to be. Ms Hang said, "I would have to ask my husband first. If I want to do something and he disagrees, I meekly ask him just once more. It works. He leaves me to decide." By so doing, she is expanding her decision-making areas. For example, she is nowadays the main decision maker in selling their maize. She says that "on rainy days, the price of maize plummets, and on sunny days the price increases sharply. My husband lets me decide when we harvest and sell the maize." Similarly, Ms Thu said, "Although I have to listen to every family member's opinions beforehand, I am still the decision maker in many areas, for example, choosing which crops to plant, how much fertilizer to be used, and the daily expenses for the family because I am the person who carries out most of these activities."

Ms Thu, 25, had an experience of her husband having an affair. While she kept silent, this experience made her stronger. She is mentally independent from her husband. She said:

> I would like to share with other women that we should not be dependent on men, but rather work for our own improvement because our own health is

> very important. I now do not expect anything from my husband. Now that I feel more independent.

Two years ago, she started her own small online business, selling clothes and imported oriental medicines. She purchases them in the district town and the neighboring province and sells them online. Her customers are from the south of Vietnam where people have high incomes. Although the profits are not very significant, they are enough for her small expenditures. Ms Thu continued,

> A pressure that Hmong married women face is that all our decisions, from what we wear to what we do, are influenced by what the others think. Before marriage, I could freely wear tank tops and shorts but now I can't because the gossip will be that I have another man. Also, we should wear makeup only for special events like weddings. Dying hair is also a taboo. I am now strong enough to do that even if I am the only one in the village. I do what I want regardless of what others, even my in-laws, say about me. If all my money came from in-laws and my husband, it would be fine for them to comment, but the money comes from my own earnings and my labour (from on-line selling of clothes). They have no right to comment on what I do. I told them not to be prejudiced against me. Many women have seen how much I care about myself and how strongly I have been trying to attain women's rights, and they try to follow my example. Otherwise they would criticise me in the women's meetings.

Thus, while women describe the life after marriage as "a bird in the cage," they are not powerless. They negotiate with their husbands and in-laws and interpret gender norms differently from what men or old women do. By so doing, they eventually expand their decision-making power, and as Ms Thu's case shows, women can change gender practices based on their own interests.

Young women, intersectional complexity and agriculture

In the study site, in addition to their domestic labor, young women often contribute a significant amount of labor to small-scale family farms, since men seasonally work as laborers outside their village and old women focus more on domestic work and childcaring. This subsection describes young men's and women's perceptions about agriculture.

Hmong young men and women marry at mid- or late teens, and the woman usually lives with the husband's parents. This situation creates very different family relationships and experiences between young men and young women. Young men are under their parents' authority both before and after the marriage. For example, Mr Quang, 23, perceives that he has little power in the household. He works for his father's farm during planting and harvesting seasons, but he and his wife do not directly receive a share of the profits, as all incomes are controlled by his parents. He thinks that his decision-making power is even lower than his wife, who at least makes decisions on everyday meals. Similarly, Mr Kien grows three pigs with his wife. However, the profits ($350 per pig) are first given to his parents, as they live together. This situation makes it difficult for young men to invest in new agricultural practices at their own risk. According to FGDs, young men perceive that a major difference between their generation and their fathers' generation in terms of agriculture is that their generation is more interested in income generating than subsistence farming. Young men are very keen on learning about cash commodities such as coffee and fruits. However, given that they have limited decision-making power, their interests and willingness are less visible, in a similar way to which women farmers were invisible in earlier agricultural research. Some young men's interests were shifted to non-agricultural sectors where they can have control of incomes. For example, Mr Dung, 29, has been seasonally working as a laborer on farms in his neighboring communes, where wages are low. Since his children are now attending school, he is thinking to go back to his previous job, a truck driver. The driving job increases burdens to his wife on looking after their children alone, as he has to be away from home for a while, but it brings a more stable income than agriculture. Mr Cuong, 26, said that if he gets a loan, he would like to invest in small business, as he is rather reluctant to engage in small-scale agriculture. Mr Nam, 25, is interested in becoming a contract farmer employed by a private company. Although it reduces autonomy, he thinks that it brings a stable income.

In contrast, young women's situations are different. They face two new types of power and authority: with in-laws and with their husbands, which affects their approaches to negotiation over family farming. Young women spend most of their time on family farms, and they are keen on improving agriculture. However, it is not easy for them to try as they want. Ms Hang recalls that in the last spring, someone was selling the planting materials of pink grapefruits for VND 30.000 ($1.5) per plant. She asked her husband if he could make a loan. Her husband said, however, that this was not something he could decide on his own (rather it needed to be discussed with his parents). She therefore had to give up purchasing the materials. Ms Thu grew free-range chickens, and they brought significant incomes, but she had to sell all of them because her parents-in-law did not like her to grow them. Her husband kept silent. As she wanted to keep family harmony, she did not argue with them. Similarly, Ms Trang said,

> I work with my parent in-laws on the field for a whole year and I never receive returns. Even all the money I earned from selling fruits has to be given to them. If I am running short of money, I have to take out loans from outside the family. If I ask them for money, they would give me money but with an attitude. My husband interfered a few times, but his father yelled at and hit my husband. We always have to spend our money for what they want.

In this way, young men and women face different challenges. The barriers of investing in agriculture for young men and women are not simply lack of money but the process of decision making and their intersectional positions within the household. Yet farming is still the most important source of food and incomes, especially for the poor households. Ms Dung expresses concern about hunger of her family and her child's weak health status. She said,

> I often think that my husband and his parents may be blaming me behind. My kid is weak and gets sick easily, we have to take him to the hospital a few times every year. My family often has financial problems and my husband takes loans. I could only bite my lip. I do not share my difficulties with anyone, even with my birth parents. . . . I often discuss with my husband what we should plant to increase incomes. We recently reduced the farm area for maize to invest more in other cash crops such as coffee and apple trees. We do our best for what we think is the best option in agriculture.

Ms Hang is aware that there are some women (from better-off households) who do small business, but she believes that working hard in her family farm is the only option for increasing her contribution to household income. She said,

> My sister, Lan, sells herbal medicine and makeup products for women online. She is literate, so she knows how to do business. She is, however, not my role model. I am aware of my limited capacity. My role model is the women who are successful in growing fruit trees. My husband only allows me to sell economically beneficial commodities for sure. He does not want to take a risk for not having sufficient returns.

Small-scale agriculture is thus still a very important source of income for young women from poor households, as unlike young men and educated women from better-off families, they have no option but working for their family farms. Given that young women often have complex intersectional power relations with their in-laws and their husbands, they do not have autonomy in family farm management, and they seek opportunities for improving livelihoods within the existing farming systems.

Gendered access to labor-saving technology

The last-subsection of this chapter focuses on gendered access to labor-saving technology. The previous study in this community finds that women sleep two hours less than men and spend on domestic work four hours more than men, while the time spent on farming is similar, around eight to nine hours for both men and women. There is no big differences in time allocation between the young (below 25) and others (above 25) (ICRAF-the Gender SRA, 2018). It is clear that women have serious time constraints in the study site. In addition, agricultural mechanization for women, namely developing labor-saving technology such as machines and tools that are suitable for women, is critical for scaling agriculture under the contexts where there is a shortage of male laborers due to men engaging in off-farm and non-farm sectors. This subsection explores to what extent women have access to labor-saving technology.

In the study site, several agricultural activities are mechanized for better-off households, such as plowing, harrowing, threshing, grinding, weeding and wood cutting. Although not all households can afford these, some machines are shared among the households in the village with small renting fees. Smaller labor-saving tools such as electric spraying tools and grass-chopping machines are available for many more households. Three key findings are described as follows.

First, the access to machines is highly gendered within the household. Some machines such as plowing, harrowing, weeding and wood-cutting machines are only used by men due to both men and women's perceptions that those machines are physically too heavy for women, or technically too difficult or dangerous for women to handle. Although some women can use them, they need men's support, while men do not need women's support to operate. In the study village, it took 30 years for women to start riding motorbikes after motorbikes were introduced to men. Currently, all young women can ride motorbikes, which significantly contributed to reducing the gender gap in mobility and increasing women's power and status. Compared to motorbikes, most agricultural machines were recently introduced in the past five years, and it may take time for women to have access to those machines. While men's use of those machines may indirectly reduce women's manual labor to some extent, many parts of women's tasks in agriculture remain manual labor. Some of these are very time consuming, such as weeding, carrying water for spraying on the hills, cooking animal feeds, picking fruits and then carrying them. Young women are very interested in small agricultural tools and machines such as fruit picking, light and small grass-cutting machines, a sprinkler and light and smaller electric spraying tools. A grass-cutting machine for women was the top priority in agricultural machines, as there is an increasing pressure from traders and consumers not to use herbicide. It was also pointed out that women need to be able to handle the machines by themselves. If they need support from men to use the machines, it rather reduces their autonomy and flexibility of their work. On the other hand, young men have limited knowledge of machines such as how much they cost and where to purchase them, although they are the main users. This is because, as discussed earlier, young men have very limited decision-making power on investing in agriculture, and most machines were purchased by their fathers.

Second, machines available in the village are only suitable for lowland fields. They are too big and heavy to carry and/or use in the steep mountain areas. Although there may be technical difficulties to develop machines which suit the uplands, this shows that the needs of ethnic-minority farmers, who are concentrated in mountain areas, are not reflected in the process of developing mechanization technology in a similar way to which women's interests were neglected. Ethnic minority farmers are interested in small and mobile tools that can be carried or used in the uplands at the affordable prices.

Third, while agricultural mechanization has been well observed, most domestic work remains manual labor. In the study site, only one household has a washing machine and 20% of households have a gas cooker, while 20% of the households still do not have a rice cooker. A water facility (tanks and pipes) helps women save time and labor for fetching water, but only 25% of the households established it. It is one of the top priorities for women. A sewing machine for clothing was owned by most households, and it is the only machine in which the ownership is entirely given to women. According to the FGDs and the photo-taking activity, women are interested in having labor-saving tools and machines more for domestic than agricultural work. Men, in contrast, do not even recognize labor-saving technology for domestic work, as they only think about agricultural tools and machines. In fact, most labor-saving machines for domestic work such as gas cooker, small water facilities and rice cooker are much cheaper than those for agricultural machines. Limited mechanization for domestic work may be associated with gendered perceptions in the priority setting and gendered and intergenerational gaps in decision-making power. As a result, most of young women's everyday tasks remain manual, while men's tasks in lowland farming are mechanized.

Discussion and conclusion

This study explored youth, in particular, young married women in a Hmong ethnic-minority village, to understand gender relations and their diverse and dynamic positions in the family and how those family relations shape everyday farming activities including their access to labor-saving technology.

The findings have shown that the behavior of young men and women is still partly shaped by norms around gender and ethnic identities that are inherited from earlier generations, but their behavior also implicitly challenges those norms. The conservative forms of identities were shown in their marriage practices in which women remain passive as shown by the stories of Ms Hang and Ms Anh. Although they had consent before marriage, their marriage processes are similar to a well-known but less common Hmong marriage practice, called "bride capture" or "wife snatching," in which a man can take a future wife without her and her parents' consent (Symonds, 2004; Jones et al., 2013). In the study site, the means of communication between boys and girls are modernized, with improved access to smartphones and internet, and social networking services allow them to explore various gender norms across the world. However, the actual Hmong women's positions in marriage remain conservative, more or less similar to their parents' generation. This agrees with a case study of Hmong adolescent girls in Ha Giang province in Vietnam (Jones et al., 2013) and also some other studies in the U.S, which prove conservative marriage views and practices among the second and third generations of Hmong migrants (e.g., Vang and Bogenschutz, 2014).

As a daughter-in-law and wife, young women experience many challenges, in particular at the beginning of their marriage. However, rather than directly challenging the existing gender norms, they often follow the norms and keep family harmony. By so doing, they eventually expand their autonomous areas, and they can then certainly challenge the typical norms. Ms Thu, for example, does not entirely follow what Hmong women should be, creating a new gender model for other young women by initiating online selling, following modern fashion and mentally being independent from her husband. Ms Hang, who did not go to school, is now significantly involved in family farm management beyond working as a family laborer that was an initial expectation from her in-laws and husband. Hmong women's implicit approaches to exercising agency agree with the literature on women's bargaining with patriarchy across the world (e.g., Kandiyoti, 1988; Kabeer, 2000; Schutte, 2014). Those types of implicit and hidden women's power cannot be captured by household surveys, which simply ask who have decision-making power. Building rapport with respondents through, for example, participatory videos was essential to have meaningful in-depth interviews, which enabled us to identify women's real capacities and opportunities within complex and dynamic family relations. Agricultural interventions have a great potential for providing opportunities to those women whose interests were less visible due to the power hierarchy of the household but who are capable to negotiate and manage.

The study also explored the processes of mechanization, with a particular focus on the processes of agricultural technology adoption at the village level. The findings show that men's agricultural activities are more mechanized than women's. Agricultural machines are male oriented in terms of color, size and weight. Agricultural work is more mechanized than domestic work. Furthermore, agricultural labor-saving technology is developed for the use of lowland farming, while in the uplands, most machines are not usable. From the perspective of the feminist theory of technology (Faulkner, 2001; Wajcman, 2010), this study agrees with interactions between gender and technology and between ethnicity and technology: marginalized social groups were not involved in the process of developing technology and household decisions on adopting technology; meanwhile, technology strengthens power to already powerful social groups, namely men and ethnic majority in this context. These processes of agricultural mechanization can partly explain the marginalization of ethnic minorities and ethnic-minority young women's positions within the household and the community, with which agricultural research has been concerned (Minot and Baulch, 2005; Friederichsen and Neef, 2010). Understanding the needs of agricultural mechanization for small-scale upland agriculture is required to transform current ongoing marginalization of ethnic minorities as well as transforming gender relations. A radical approach is needed to transform the processes – rather than increasing women's access to the existing male-oriented technology and adjusting machines for lowland farming to the uplands, we need to start identifying needs and designs for women, in particular, for ethnic-minority young women as well as men.

Lastly, this study confirms that multi-dimensional and dynamic gendered power relations play a central role in the family members' security and well-being (e.g., Guyer and Peters, 1987; Hart, 1997; Carr, 2005). As such, listening to young women's as well as young men's voices is so important for providing appropriate agricultural support to them, but their voices are less likely to be captured in the simple household survey and the village group meetings, where power based on gender and age play out. In this respect, this study confirms that participatory approaches such as taking photos and videos with respondents are powerful means of communication for listening to those who are in the marginalized positions. Understanding complex power dynamic with gender, generation and ethnicity is an entry point for more equitable and inclusive agricultural development.

Acknowledgment

We thank Hmong young men and women participants and the local authority for their support in the youth's participation in this study, including film making and organizing the screening event. We also thank ICRAF Vietnam (FST/2016/152) for their institutional support. The study was funded by ACIAR (AGR/2017/008) and CGIAR research program on Roots, Tubers and Bananas (RTB). We appreciate valuable comments on our draft paper from Gordon Prain (CIP) and Ann Tickamyer and Carolyn Sachs (Penn State University).

References

Braun, V. and Clarke, V. (2006). Using thematic analysis in psychology. *Qualitative Research in Psychology*, 3(2), pp. 77–101.

Bonnin, C. and Turner, S. (2014). "A good wife stays home": Gendered negotiations over state agricultural programmes, upland Vietnam. *Gender, Place & Culture*, 21(10), pp. 1302–1320.

Butler, J. (1990). *Gender trouble: Feminism and the subversion of identity*. New York: Routledge.

Carr, E.R. (2005). Development and the household: Missing the point? *GeoJournal*, 62(1–2), pp. 71–83.

Connell, R. (2009). *Gender in world perspective*. 2nd ed. London: Polity Press.

Doutriaux, S., Geisler, C. and Shively, G. (2008). Competing for coffee space: Development-induced displacement in the central highlands of Vietnam. *Rural Sociology*, 73(4), pp. 528–554.

Elmhirst, R. (2011). Migrant pathways to resource access in Lampung's political forest: Gender, citizenship and creative conjugality. *Geoforum*, 42(2), pp. 173–183.

Elmhirst, R., Siscawati, M., Basnett, B.S. and Ekowati, D. (2017). Gender and generation in engagements with oil palm in East Kalimantan, Indonesia: Insights from feminist political ecology. *The Journal of Peasant Studies*, 44(6), pp. 1135–1157.

Faulkner, W. (2001). The technology question in feminism: A view from feminist technology studies. *Women's Studies International Forum*, 24(1), pp. 79–95.

Friederichsen, R. and Neef, A. (2010). Variations of late socialist development: Integration and marginalization in the northern uplands of Vietnam and Laos. *The European Journal of Development Research*, 22(4), pp. 564–581.

Guyer, J.I. and Peters, P.E. (1987). Introduction. *Development and Change*, 18(2), pp. 197–214.

Hart, G. (1997). From "rotten wives" to "good mothers": Household models and the limits of economism. *IDS Bulletin*, 28(3), pp. 14–25.

ICRAF-the Gender SRA. (2018). *Understanding opportunities and challenges in agricultural development through the gender lens: A case study in a H'mong community in Dien Bien*. A report submitted to ACIAR. Available at: https://genderinagr.files.wordpress.com/2018/04/a-joint-report-gender-final.pdf.

Jackson, C. (1998). Gender, irrigation, and environment: Arguing for agency. *Agriculture and Human Values*, 15, pp. 313–324.

Jones, N., Presler-Marshall, E. and Tran Thi Van Anh. (2013). *Early marriage among Viet Nam's Hmong: How unevenly changing gender norms limit Hmong adolescent girls' options in marriage and life*. Country Report, p. 113. London: Overseas Development Institute.

Kabeer, N. (2000). *The power to choose*. London: Verso.

Kandiyoti, D. (1988). Bargaining with patriarchy. *Gender & Society*, 2(3), pp. 274–290.

Lindeborg, A.K. (2012). *Where gendered spaces bend: The rubber phenomenon in Northern Laos*. Doctoral dissertation, Uppsala University, Sweden.

Minot, N. and Baulch, B. (2005). Spatial patterns of poverty in Vietnam and their implications for policy. *Food Policy*, 30(5–6), pp. 461–475.

Nightingale, A. (2006). The nature of gender: Work, gender, and environment. *Environment and Planning D: Society and Space*, 24(2), pp. 165–185.

Resurreccion, B.P. (2006). Gender, identity and agency in Philippine upland development. *Development and Change*, 37(2), pp. 375–400.

Schütte, S. (2014). Living with patriarchy and poverty: Women's agency and the spatialities of gender relations in Afghanistan. *Gender, Place & Culture*, 21(9), pp. 1176–1192.

Scott, J.C. (2009). *The art of not being governed: An anarchist history of Upland Southeast Asia*. Singapore: Nus Press.

Smith, C.P. (1992). *Motivation and personality: Handbook of thematic content analysis*. Cambridge: Cambridge University Press.

Suarez-Ortega, M. (2013). Performance, reflexivity, and learning through biographical-narrative research. *Qualitative Inquiry*, 19(3), pp. 189–200.

Symonds, P.V. (2004). *Calling in the soul: Gender and the cycle of life in a Hmong village*. Seattle: University of Washington Press.

Turner, S. (2012). "Forever Hmong": Ethnic minority livelihoods and agrarian transition in upland Northern Vietnam. *The Professional Geographer*, 64(4), pp. 540–553.

Vang, P.D. and Bogenschutz, M. (2014). Teenage marriage, and the socioeconomic status of Hmong women. *International Migration*, 52(3), pp. 144–159.

Wajcman, J. (2010). Feminist theories of technology. *Cambridge Journal of Economics*, 34(1), pp. 143–152.

West, C. and Zimmerman, D.H. (1987). Doing gender. *Gender & Society*, 1(2), pp. 125–151.

Whitehead, A. (1981). I am hungry, mum: The policies of domestic bargaining. In: K. Young, C. Wolkowitz and R. McCullagh, eds., *Of marriage and the market*. London: CSE Books.

World Bank. (2009). *Country social analysis: Ethnicity and development in Vietnam*. Washington, DC: World Bank.

Part 4

Shifting gender relations in agriculture, fisheries and livestock

8 Defeminizing effect

How improved dairy technology adoption affected women's and men's time allocation and milk income share in Ethiopia

Birhanu Megersa Lenjiso

Introduction

In the Ethiopian traditional and subsistence oriented dairying system, dairy is a woman's business. The women undertake almost all the dairy activities. They manage the cows and calves; milk the cows, process milk into cottage butter and cheese both for household consumption and for sale in the local market. The income generated from butter and cheese sold in local market is an important income for rural women in Ethiopia. They fully control and spent this income, and it is a taboo for men to be involved in the local dairy business (Feleke, 2003; Lenjiso et al., 2015).

Moreover, selling raw milk has been a social taboo in Ethiopia, especially among the Oromo society. Taking milk and milk utensils outside of the farm compound is believed to cause damage to the entire cattle farm (Lenjiso, 2016). However, there have been efforts to modernize the Ethiopian dairy sector for over half a century against this tradition. These efforts were started in 1947 when the Imperial government received 300 Friesian and Brown Swiss dairy cows from the United Nation Relief and Rehabilitation Administration as donation and established a small milk processing unit in the Shola area. Following this, state and private dairy farms started to mushroom in the country. In 1975, the socialist government nationalized most private farms, and the rest were forced to become members of cooperatives to get dairy services. That is believed to have undermined dairy development initiatives during the socialist regime in the country.

The current government of Ethiopia has been implementing the growth and transformation plan since 2010. This plan emphasizes adoption of productivity boosting technologies and commercialization of the entire agricultural sector as a major pathway from a semi-subsistence agrarian society to a more diversified and food-secure Ethiopia (MoFED, 2010). The plan specifically encourages liberalization of dairy markets, introduction of improved cattle breeds, improved fodder and new cattle management skills. This is believed to increase smallholder dairy farmer's productivity, their market participation, and income and thus improve food security and welfare of the producers. It is also believed that this strategy will offer high return to input use and empower women and youth. Moreover, the strategy is suitable for adoption by the smallholders, who will continue to dominate the Ethiopian agriculture sector in general and dairy in particular.

Advocates of the technology-based development and commercialization agenda argue that technology adoption and commercialization is an indispensable pathway toward economic growth, development and food security among agricultural-dependent societies (Timmer, 1997; Tangka et al., 1999). However, Endeley (2001) argues that development policy makers need to be critical in analyzing cultural contexts and question the likelihood that development interventions will change the gendered power relations. This is because access to income at household level is not necessarily linked with achieving control over it or the right to participate in social, economic or political decision making in the household or outside.

According to Caraway (2006) and Kucera and Tejani (2014), sectors that use more labor-intensive production methods tend to employ a higher share of women workers while sectors that use capital-intensive production methods employ men. Therefore, shifting labor-intensive sectors like Ethiopian dairy into capital-intensive production systems by introducing productivity boosting improved technologies may defeminize the production sector by pushing women out of the sector or by diminishing their control over the benefits. With the gendered subsistence versus cash commodity division in Africa, transforming subsistence commodity into cash commodity may also imply defeminization in a sense that men tend to take over those subsistence commodities once they become commercialized and start to generate income. This is depriving for women's control over benefits, while they contribute much of the labor.

In Ethiopia, dairy technology adoption and the subsequent commercialization of milk are basically inspired by the objective of increasing household income. Hence, smallholder farmers in peri-urban areas of Addis Ababa are producing milk for sale and young people are increasingly engaging in milk collection and transportation to the vibrant Addis Ababa milk market. The social taboo associated with milk selling in the area does not apply anymore partially because farmers are milking the giant outsider (foreigner) cow on one hand and because of the changing belief system on the other hand. Although there are still farmers who do not sell milk and others sell only milk from crossbred cows, milk sale is a base of livelihood in peri-urban areas of Addis Ababa. This is among the changes that followed the introduction of new dairy technologies in the study area. There are also associated shifts in dairy roles and responsibilities and intra-household income control structure. However, these issues are less studied. Hence, this paper aims to analyze men's and women's dairy time allocation, changes in their role and responsibilities and their share of the dairy income by comparing smallholder dairy farmers who rely on traditional dairy cow breeds with those who adopted the improved cattle breeds with new fodder technologies and new cattle management skills.

The rest of this paper is structured as follows. In section 2, a brief literature review is presented, section 3 presents methods and materials, section 4 presents the empirical results of the study, and in section 5, the major findings are discussed.

Literature

The context

Ethiopia is the world of smallholders. Subsistence agriculture makes the backbone of the country's economy. Smallholder farmers cultivate 95% of the total crop land and produce more than 90% of total agricultural output (Mahelet, 2007). The livestock sector makes the lifeline of about 65% of the Ethiopian population. It contributes about 12% to the total economy, 33% to the agricultural gross domestic product and 12–15% of the export earnings.

Given its livestock population and the favorable climate for production, Ethiopia has a high potential for dairy development. The country has the largest cattle population in Africa. Ethiopia has also a relatively low animal disease incidence and a favorable climate for cattle rearing (Chagwiza, 2014). However, the dairy sector is largely subsistence oriented with low milk production (4 billion liters of milk per annum) and low consumption (26.6kg/year/person). There have been efforts to modernize the dairy sector since the 1940s. Governmental and non-governmental organizations have made efforts to introduce new technologies that boost productivity and market orientation of the dairy farmers. This attempt was started by the imperial government, who received 300 Friesian and Brown Swiss dairy cows from the United Nation Relief and Rehabilitation Administration in 1947. A small milk processing company was established in Shola to support the commercial dairy production. In 1975, the socialist government nationalized land and dairy farms and started the formation of peasant associations and producer cooperatives. However, the low price of dairy products and the strict government control of cooperatives negatively affected individual dairy farmers and hindered dairy development in the country during the socialist era (Feleke, 2003; Staal et al., 2008).

In the 1990s, Ethiopia came to adopt a relatively liberal market economy. Since then, the government has stimulated dairy transformation from a subsistence oriented production system into a market-oriented system that is characterized by new technology adoption (MoFED, 2010). Although the dairy sector still remains by and large traditional in the country, there is increasing adoption of new technologies in peri-urban areas. Smallholder milk market participation is also increasing from time to time against the norm that makes milk sale a social taboo. The growth and transformation plan that is in second phase of its five years now, encourage agricultural transformation in terms of technology adoption and commercialization for improving the welfare of every member of the society.

Ethiopian women's role in dairy operation

According to EADD (2009) and Mutinda (2011), women have a long-standing traditional role in livestock production in the East Africa region. Similarly, women have always played a central role in the Ethiopian dairy industry. Traditional subsistence-oriented dairy was a woman's business. They managed the cows, cleaned their sheds, milked the cows and processed the milk into cottage butter and cheese. They also sell the market surplus butter and cheese in the local market. The income generated in this way was controlled by the women and generally used for household maintenance. Although they participate in milk production, collection, processing and marketing locally, women's participation in the formal dairying value chain is often low and passive, as they tend to contribute much at the production level but miss out in subsequent more profitable levels such as accessing membership in farmer organizations (EADD, 2009; FAO, 2010). The fact that co-operative membership is household based in Ethiopia gives the head of the household (often men) access to extension activities, training, information, leadership and decision making, access to improved technologies, marketing and financial services.

Moreover, female farmers' preferences in market-led approaches differ substantially from their male counterparts (FAO, 2010). Development approaches that does not take into account these gender needs and preferences faces the risk of reinforcing, reproducing and/or maintaining gender inequalities, often propagated by prevailing socioeconomic and institutional frameworks, which determine who does what, who gets what and who passes which decision (Mutinda, 2011).

Research by EADD (2009), Mutinda (2011) and FAO (2010) agree that women are represented disproportionately in low value chains, and the lower-value nodes within them and men tend to dominate functions with relatively high barriers to entry and correspondingly greater returns. Observation studies from Ethiopia show that women's intra-household workload increased with shifting cattle management practices in milk market–integrated households (Lenjiso et al., 2015).

There is a large body of evidence on intra-household time allocation in production activities in developing countries. One of the major activities of the household, as an institution that govern the time allocation of its members is to prioritize, assign members to different activities and reward them based on their contributions. In this, sense, technology adoption and milk commercialization could affect household time allocation, as a market-oriented production system could aim to create income-generating opportunities from previously subsistence commodities (Ilahi, 2000). Intuitively, individuals allocate labor or time in certain productive activities only if they get compensated well for doing that. However, it is also known that not all household members have equal say on their own time and thus decide where to invest their time based on the benefit they generate.

Gender, technology adoption and control over benefits

Most technology adoption initiatives in Africa take the existing gender relations for granted. However, the exiting gender relations directly influence the likelihood for adopting technologies, and technologies tend to influence existing gender relations (Doss, 2001). Technologies are gendered (1) because they are generally perceived as males' domain in developing countries and (2) because technology adoptions are influenced by the existing structure for accessing information and productive resources and the resulting benefits. As access to productive assets and control over the benefit is gendered, technology adoption processes cannot be gender neutral.

Moreover, most agricultural technology adoption aims to increase households' income, meet demand for agricultural product and enhance food security. However, the income generated is not often shared equally between husband and wife. A gender-differentiated study on the income impact of dairy technology adoption in Ethiopia indicated that men and women in households with crossbreed cows had 14 and 4 times higher dairy incomes than men and women in households with indigenous-breed cows, respectively (Tangka et al., 1999). Thus, although both women and men benefit from dairy technology adoption and market-oriented dairying, men's cash incomes rise significantly more as a result of dairy technology adoption and commercialization compared to those of women. This also entails that dairy technology adoption and the subsequent milk market integration can have a potential for increasing intra-household income inequality between men and women.

Coles and Mitchell (2011) and Tangka et al. (1999) recommended examining the impact of the introduction of new technologies and market-based policies on women's control over the income generated, at least the income that they traditionally used to control. The intra-household changes in income control with transition from traditional to market orientation and its effects on women's role and responsibilities is thus a critical area in gender and agricultural commercialization studies. FAO (2010) indicated that commercial dairying, besides worsening women's workload, is likely to erode women's traditional control over milk products, especially in transitional pastoral areas. Without taking into account household gender relations, no intervention can lead to meaningful and sustainable empowerment (EADD, 2009; FAO, 2010).

In general, there is a large body of knowledge on how technology adoption boosts productivity and household income and thus improves household welfare. However, how these linkages influence and are influenced by the existing gender norms is less known. The major aim of this chapter is therefore to contribute to the literature on how agricultural technology adoption affects the gender dynamics through its effect on men's and women's time allocation and their control over its benefits. Figure 8.1 presents the potential linkages between gender, dairy technology adoption and control over benefits.

Methods and materials

Site selection and sampling

This study was conducted in Selale, situated in the Oromia national regional state of Ethiopia. Oromia is the largest state of the nine regional states in Ethiopia. It represents one-third of the total area of the country and inhabits close to 40% of the country's total population (CSA, 2008). The region is classified

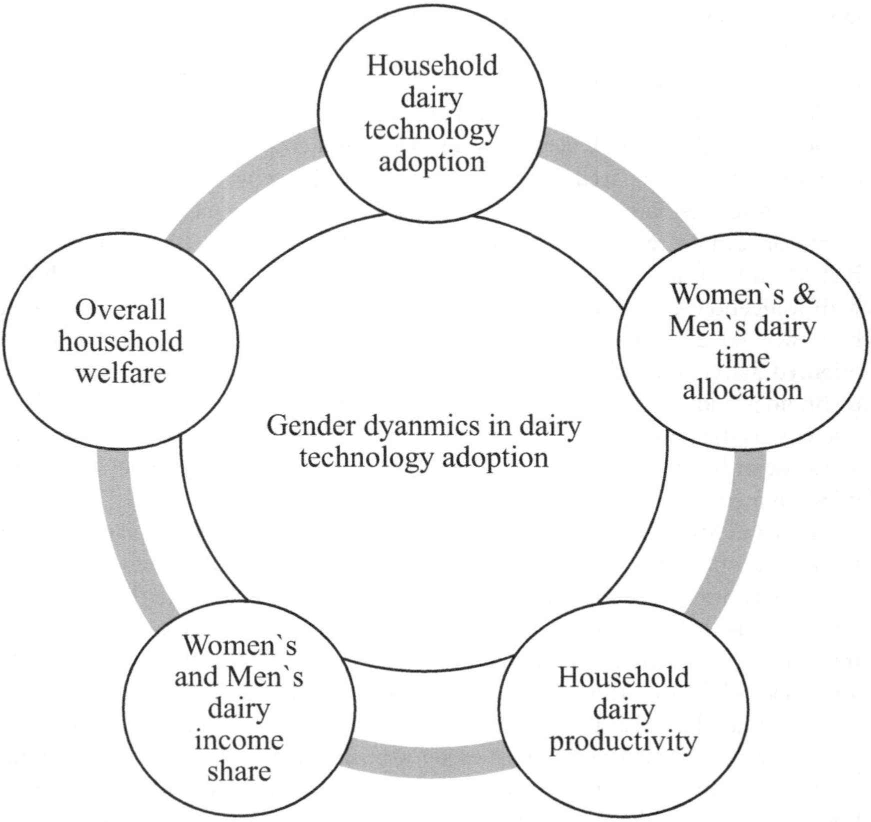

Figure 8.1 Links between gender, dairy technology adoption and household welfare

into 18 zones. Selale is one of the 18 zones in Oromia, known for its high dairy production. It was purposefully selected for this study because of its long-standing dairy tradition, farmers' dairy technology adoption rate, dairy supply to the national market, and dairy infrastructures that have facilitated dairy production and marketing.

Selale is relatively homogeneous in terms of ethnicity and religion. The Oromo, the largest ethnic group in Ethiopia, makes up about 95% of the population, and most of them are followers of the Orthodox Christian church. The crop–livestock mixed-farming system and milk market coverage is similar across Selale. For this research, a sample of four primary dairy cooperatives was randomly selected from the list of 22 primary dairy cooperatives that are active members of the Selale dairy co-operative union (SDCU), the largest and the only formal milk buyer in the area. The four *kebeles* where the sample dairy cooperatives are located (one dairy cooperative being in a *kebele*) were taken as a sample *kebele* for the study. By employing stratified sampling techniques, 296 farm households were selected from the four kebeles proportional to the size of their population. The survey sample included 139 dairy farmers who own exotic/crossbreed cows and adopted new fodder technologies and 157 traditional dairy farmers who own indigenous-breed cows and rely on free grazing. Government officials working in agricultural offices and development agents were included in the key informant interviews.

Study approach, data collection and analysis

The study is based on a mixed-methods approach in which both quantitative and qualitative techniques have been used in the form of triangulation and with the intention of backing numerical information with qualitative aspects. As the techniques are different, having their own strengths and weaknesses, the combination of these techniques is believed necessary to enrich the research by triangulating findings of one method against the other; to benefit from the strengths of each method and to compensate for the weaknesses; and to establish confidence and obtain trustworthiness of the research findings. The key informant interviews were believed to put some flesh on household survey skeletons. Accordingly, a structured questionnaire was developed and distributed to 296 dairy farmers to generate quantitative information on their experiences with dairy technology adoption, dairy farming and marketing, their daily dairy time allocation and their share of daily dairy income. We also performed key informant interviews to gain an insight into the change in the farming system and its effects on the status of women in the dairy business.

Analytical approach

This study aims to measure the average effect of household dairy technology adoption on men's and women's time use and their share of the dairy income, which will be called the *average treatment effect on treated households* (ATT) (Heinrich et al., 2010; Austin, 2011). The treatment is household dairy technology adoption. Households that adopted improved cattle breeds and improved fodder technologies are considered technology adopters. Households located in the same kebele as the adopter households but that rely on traditional breeds and free grazing are non-adopters. We used propensity scores to match adopters and non-adopter households on selected baseline characteristics to ensure that the effect on the outcome variable was the result of difference in dairy technology adoption.

It is obvious that observational studies that aim to establish causal relationships face several methodological challenges. The first of these challenges is lack of the counterfactuals. As there are no baseline data in many observational studies and the treatment group cannot undergo and not undergo the treatment at the same time, the question "what would have happened to the treated group without the treatment" cannot be answered. Although constructing a control group can help to address this problem, control groups may not have the same baseline characteristics as the treatment group. The difference in baseline characteristics may affect both treatment uptake and outcome, which can lead to selection bias.

Recently, there has been growing interest in methods based on propensity scores to reduce these selection biases (Austin, 2011). According to Heinrich et al. (2010), the effect of a treatment for an individual i, noted δ_i, can be defined as the difference between the potential outcome in case of treatment and the potential outcome in absence of the treatment: $\delta_i = Y_{1i} - Y_{0i}$. The major aim of evaluation, according to the authors, is to estimate the mean effect of the intervention, obtained by averaging the effect across all the individuals in the population. This parameter is known as Average Treatment Effect (ATE), which is given as; $ATE = E(\delta) = E\ Y(Y_1 - Y_0)$ where $E(\delta)$ represents the average (or expected value).

The other variable of interest according to Heinrich et al. (2010) is the ATT discussed above, which measures the impact of the program on those individuals who participated in the intervention: $ATT = E\ (Y_1 - Y_0 | D = 1)$. Finally, the Average Treatment Effect on the Untreated (ATU) measures the impact that the program would have had on those who did not participate: $ATU = E\ (Y_1 - Y_0 | D = 0)$. However, this depends on counterfactual outcomes and is unobservable. Using the fact that the average of a difference is the difference of the averages, the ATT can be rewritten as: $ATT = E\ (Y_1 | D = 1) - E\ (Y_0\ | D = 1)$. $E\ (Y_0 | D = 1)$ is the average outcome that the treated individuals would have obtained in the absence of treatment, which is not observable. However, we do observe the term $E\ (Y_0\ | D = 0)$, that is, the value of Y_0 for the untreated individuals.

Thus, we can calculate: $\Delta = E\ (Y_1 | D = 1) - E\ (Y_0\ | D = 0)$ By adding and subtracting the term $E(Y_0 | D = 1)$: $\Delta = E\ (Y_1 | D = 1) - E\ (Y_0 | D = 1) + E\ (Y_0 | D = 1) - E\ (Y_0 | D = 0)$, $\Delta = ATT + E(Y_0 | D = 1) - E(Y_0 | D = 0)$, $\Delta = ATT + SB$. The second term, SB, is the selection bias: the difference between the counterfactual for treated individuals and the observed outcome for the untreated individuals. When this term is equal to zero, then the ATT can be estimated by the difference between the mean observed outcomes for treated and untreated groups: $ATT = E\ (Y | D = 1) - E\ (Y | D = 1)$.

Heinrich et al. (2010) identified two main assumptions for propensity score matching. The first is the conditional independence assumption (CIA) that supposes there is a set of X covariates observable to the researcher such that after controlling for these covariates, the potential outcomes are independent of the treatment status. The second is the common support condition, which assumes that for each value of X, there is a positive probability of being both treated and untreated. Based on these assumptions, propensity score matching can make selection bias (SB) zero, and ATT is an unbiased estimator of the impact of the intervention on the treated.

Based on these assumptions, this study follows two steps of analyses. In the first step, it identifies the major determinants of household dairy technology adoption and the households are matched on those determinants. In the matching procedure, it was identified that here are households both from the adopter and non-adopter groups on the common support area. The distribution of the estimated propensity scores is presented in Figure 8.2.

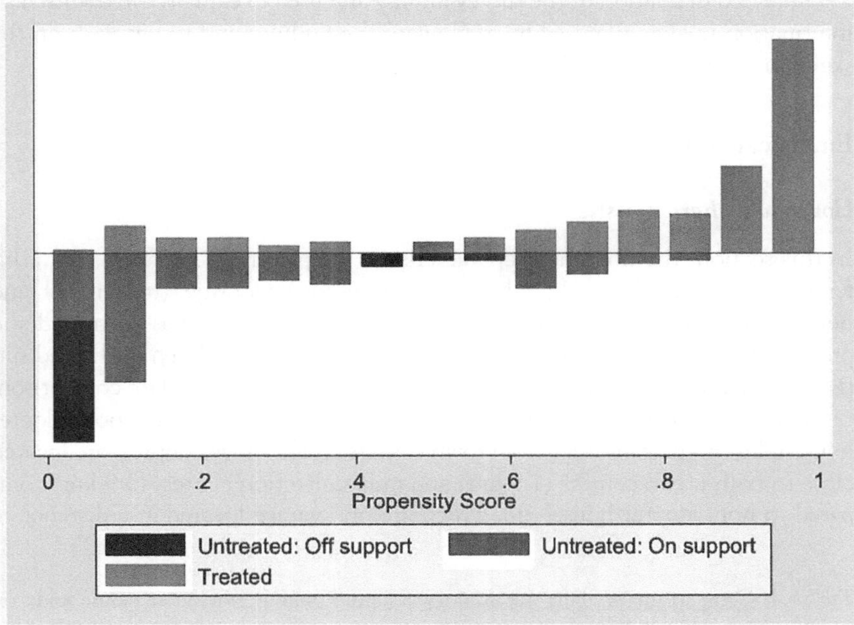

Figure 8.2 Distribution of the propensity scores from dairy technology adopters (treatment group) and non-adopters (control group)

In the second step, two matching algorithms; nearest neighbor (NN) matching "with replacement" and kernel matching were used to estimate ATT, the mean difference in men's and women's dairy time allocation and their milk income share for dairy technology adopter and non-adopter households. According to De Hoop (2012), using these two matching algorithms provides a natural robustness check to guard against the disadvantages of the two matching algorithms. In the NN matching, the individual from the control households is chosen as a matching partner for a treated individual that is closest in terms of propensity score (Caliendo and Kopeinig, 2005). "Matching with replacement" allows an untreated individual to be used more than once as a case. Kernel matching is a non-parametric matching estimator that uses weighted averages of all individuals in the control group to construct the counterfactual outcome (Caliendo and Kopeinig, 2005). Therefore, everyone in adopter households is matched with the entire sample of individuals in non-adopter households. This technique uses more information from the control household and can lower variance. However, it might include observations that are bad matches. Thus, the common support condition has been imposed to minimize this drawback and to improve the robustness of the results. The standard errors have been computed using 100 bootstrap replications.

The qualitative information gathered through the key informant interviews was analyzed in relation to the survey finding and used to explain the results. The information was mainly used to triangulate the findings and to put flesh on the skeleton of the survey results.

Empirical results

Household characteristics

In this section, the household characteristics are presented based on mean difference on selected variables between technology-adopting (treatment) and non-adopting (control) households. For this purpose, a *t*-test has been used and presented in Table 8.1. The comparison of the two household types showed statistically significant differences on 18 of the 23 variables included for comparison.

There is a statistically significant difference between adopters and non-adopter households on their distance from town centers. Technology adopters are located close to both urban centers (1.9 km) and milk collection centers (3.9 km) compared to non-adopter household, which are on average located at a distance of

Table 8.1 Comparison of dairy technology adopters and non-adopter households on selected variables

Comparison Variables	*Mean adopters*	*S.E*	*Mean non-adopters*	*S.E*	*T-Stat*
Distance from town	1.8842	.150	3.234	.154	−6.244***
Distance from milk market	3.911	.187	4.606	.155	−2.890***
Respondent age	44.741	.801	45.420	.725	0.630
Respondent education	.806	.034	.752	.034	1.115
Total household size	7.467	.184	6.734	.159	3.037***
Total male size	3.7843	.122	3.758	.106	0.1624
Total female size	3.683	.111	2.936	.092	5.211***
Total household land size	4.489	.204	3.885	.157	2.378***
Total household cow size	4.885	.138	3.949	.1041	5.481***
Indigenous cow size	2.179	.142	3.287	.110	−6.227***
Crossbreed cow size	2.683	.111	.574	.051	17.923***
Fodder adoption	1.000	0.000	.140	.0278	29.106***
Dairy experiences	18.201	.873	18.516	.589	0.304
Dairy related trainings	.6978	.039	.477	.039	3.916***
Co-op membership	.633	.041	.223	.033	7.829****
Total milk production	9.905	.605	5.047	.184	8.071***
Formal Milk sales	6.737	.627	1.166	.171	9.020***
Milk processed at home	2.083	.192	2.818	.164	−2.929***
Milk consumed at home	1.144	.113	1.0763	.091	0.460
Women's dairy time	9.896	.205	6.895	.1848	10.885***
Men's dairy time	4.392	.177	3.598	.145	3.496***
Women's dairy income share	16.662	1.5391	22.853	1.382	−3.001***
Men's dairy income share	53.899	5.019	9.121	1.363	9.065***

***$p < 0.01$ **$p < 0.05$ * $p < 0.1$ *Number of observations 296 (139 dairy technology adopters and 157 non-adopters)*

3.2 km from urban centers and 4.6 km away from milk collection centers. There is no statistically significant difference between the ages of the respondents from the two household types. Similarly, literacy rate among respondents from technology-adopting and non-adopting households does not show statistically significant difference, but technology adopter households have significantly larger household size (7.4) and significantly more female members per household (3.7) compared to non-adopters (6.7 and 2.9, respectively).

Technology-adopting households have significantly larger land size (4.48 hectares) compared to non-adopters (3.9 hectares). They also have significantly more cows (4.88) and crossbreed cows (2.68) compared to non-adopters, who have 3.9 for total cows and 0.573 crossbreed cows. On the contrary, adopter households have significantly fewer indigenous-breed cows (2.17) compared to non-adopters (3.28).

All technology-adopter households have adopted some sort of fodder technology, while only 14% of the non-adopters adopted fodder technology. The households do not differ on their dairy experiences. About 70% of the dairy technology adopters have taken some sort of dairy-related training while only 48% of the non-adopters are exposed to dairy-related training. Similarly, 63.3% of adopter households are members of dairy cooperatives, while only 22% of the non-adopter households are members of dairy cooperatives.

Total household milk production per day is significantly higher for adopter households (9.9 liters) compared to non-adopters (5.04 liters). Milk is either sold raw to the formal milk market (cooperatives, milk collectors and hotels/cafeterias), consumed at home or processed into cottage butter and cheese at the household level for both consumption and sale in the local market. The volume of milk sale to the formal milk market is significantly higher among adopter households (6.7 liters compared to 1.16 liters among non-adopters), while the average volume of milk processed into butter per day is significantly lower among adopter households (2.08 liters compared to 2.89 among non-adopters). Despite the significant production difference, there is no significant difference in milk consumption at home between the two household types.

The two households also significantly differ on men's and women's time allocation on dairying. Women invest generally more time per day on dairy activities compared to men. They allocated 9.9 hours to dairy-related activities per day in technology-adopter households and 6.9 hours per day in non-adopter households. Men allocate 4.39 and 3.59 hours respectively. Although both men and women allocate significantly more time on dairy in adopter households per day, women's time increased more significantly compared to men's.

The difference also holds for intra-household milk income sharing. Income from formal milk sales is received by men, while income from processed milk (butter and cheese) is received and controlled by women in the area. In line with this, our key informant interview participant provided the information that strengthen this income-sharing tradition:

> the idea of selling raw milk was debated and challenged in the community and among household members in the study area. Traditionally, dairy income was the domain of women and men's involvement was considered a taboo.

> However, with improved dairy technology adoption formal milk sales is increasing and the households who sell milk to private collectors, processing companies and cooperatives have to sign a contract and receive milk income twice a month through a registered head of the households (men).
>
> (Yosef, male, agriculture expert)

Based on this information, we compared women's and men's intra-household milk income share. Accordingly, men's income share increased significantly with dairy technology adoption, while women's income share reduced significantly. In technology-adopting households, men receive 53.9 birr per day from the formal milk market, while women receive 16.7 birr. On the contrary, women receive 22.8 birr per day from milk, while men receive 9.12 birr in non-adopter households.

Based on these evidences, women's dairy time increased by 43.5%, while their milk income share (control) reduced by 26.7% with dairy technology adoption. On the other hand, men's dairy time increased by 22%, while their milk income share (control) increased by 491% in adopter households compared to non-adopter households.

The defeminizing effect of dairy technology adoption

Effect on women's and men's daily time use

The average effect of household dairy technology (fodder and dairy breed) adoption on women's and men's time allocation on dairy-related activities is presented in Table 8.2. As the result shows, women in dairy technology adopter households invest significantly more time on dairying compared to women from non-adopter households. They roughly invest 3 hours more in dairying per day compared to women from non-adopter households.

Similarly, men from dairy technology adopter households invest significantly more time in dairy-related activities compared to men from non-adopting households. They invested 1.5 hours more on dairy compared to men from non-adopter households. However, the difference is only half of the difference between women from adopter and non-adopter households.

Table 8.2 Women's and men's daily time allocation in dairying

Variable	*NN Matching*			*Kernel Matching*		
	Mean Differ	*S.E.*	*T-stat*	*Mean Differ*	*S.E.*	*T-stat*
Women's daily dairy Time	2.88	0.439	6.57***	2.67	0.296	8.99***
Men's daily diary Time	1.16	0.424	2.73***	1.27	0.324	3.94***

Note: ***$p < 0.01$ **$p < 0.05$ * $p < 0.1$ *Number of observations 296, observation on comsup = 261, numbers of treated = 139 and controls = 122.*

Our key informants added to this fact that women are also involved in many other agricultural and household activities, which makes their workload heavy.

> In Ethiopia women widely participate in cooking, carrying for children, fetching water and firewood, cleaning the home, clothes and barn, dung collection and dung cake preparation, cow milking and milk processing, preparing concentrates for milking cows and calves and marketing dairy products. They also widely participate in crop production activities and other community social affairs that depends on their labor. In dairy women have significant contribution but their participation and contributions are confined to production stage and their participation decrease as we move downstream in the value chain.
>
> (Abebech, Development Agent)

Yosef also added on a similar line. He stressed that there is a shift in cattle management in the study area. With declining farm size and increasing government restrictions on free grazing, smallholder farm households are increasingly adopting improved dairy breeds that feed on concentrates and improved fodder technology. Since that most often happens in the backyard, feeding and watering cows falls within the domestic chores that women are traditionally responsible for. But men are also increasingly taking part in traditional women's dairy roles and responsibilities due to the heavy dependency on milk production in the area.

Effect on women's and men's milk income share

The average effect of household dairy technology adoption on men's and women's daily dairy income share is presented in Table 8.3. The result indicates that women's dairy income share showed negative in technology adopter households compared to non-adopters. Although the difference is not statistically significant, it shows that their dairy income share is not increasing compared to women's

Table 8.3 Effect on women's and men's daily milk income share

Variable	*NN Matching*			*Kernel Matching*		
	Mean Differ	*S.E.*	*T-stat*	*Mean Differ*	*S.E.*	*T-stat*
Women's daily milk income share	−10.04	9.09	−1.10	−3.22	6.80	−0.47
Men's daily milk income share	13.57	7.27	1.98**	15.06	6.36	2.37***

Note: ***$p < 0.01$ **$p < 0.05$ * $p < 0.1$ *Number of observation 296, observation on comsup = 261, numbers of treated = 139 and controls = 122.*

dairy income share in non-adopter households. They earn between 3 and 10 birr less from dairying daily compared to women from non-adopter households.

On the contrary, men's daily milk income share has significantly increased with dairy technology adoption. Men from dairy technology adopter households earn between 13 and 15 birr more from dairy per day compared to men from non-adopter households.

It is clear that the gender relations that govern society at the higher level and family structure at the lower level also govern women's and men's access to and control over dairy resources and services and their control and share of the benefits. In relation to this, the key informants stressed that men and women should have a fair share of benefits from increasing dairy production if we have to sustain the increasing dairy production and meet the increasing demand for milk in urban and peri-urban areas.

> The demand for milk is predicted to increase in our country. Dairy farmers, especially female farmer's access to and secure rights over productive resources like land, labor, livestock and other dairy technology and services is a prerequisite for improving dairy production and also to improve the status of women farmers in the country. In more traditional dairy production, given the small amount of milk production men don't bother their women. But with increasing dairy production and marketing men's involvement is increasing and that is already causing a lot of challenges at household level and beyond.
>
> (Abera, Livestock and Dairy expert)

Moreover, the fact that dairy income is primarily used for home consumption and that the household needs are always greater than the milk income that women generate from the local market, the declining share of women's daily milk income could have a serious implication for household welfare.

Discussion

This study reports household survey and key informant interview results on the effect of dairy technology (improved dairy breeds and fodder) adoption on women's and men's daily time allocation in dairying and on men's and women's daily milk income share (control). It is widely reported that technological innovation with traditionally women-controlled commodities and its commercialization leads to shifts in intra-household income control from women to men (Udry, 1996; Tangka et al., 1999; Endeley, 2001; Njuki et al., 2011). There is also evidence that some of those innovations increase women's workload, while their control over benefit either declines or doesn't show significant difference. Over the last 15 years, the Ethiopian government has been implementing growth and transformation plans that aim to transform the agricultural sector in general and the dairy sector in particular from subsistence to more commercial production systems (MoFED, 2010). The assumption behind the plan is that transforming agriculture from subsistence to more market-oriented production systems will make farmers competitive, increase income at household level, meet demand for agricultural product and thus solve the food security challenge in the country and end poverty.

The efforts to transform dairy in Ethiopia has a long history. Over the last 70 years, new dairy technology (i.e., improved cattle breeds, new fodder technology, chilling plants and better cattle management skills) adoption has been in trial, but the adoption rate is generally poor. Farmers in the study area in general and women farmers in particular show interest in productivity-boosting dairy technologies, but their interests are challenged by weak access to and control over resources. Moreover, the fact that technologies attract the attention of men has forced women to remain in the informal, traditional dairy production system. In the peri-urban areas, one can generally say the dairy technology adoption and the subsequent milk commercialization have achieved their policy objectives by increasing household income. Technology adopters, who are located close to urban centers and milk collection centers, also have large land and family size and better dairy-related trainings and are members of dairy cooperatives. They also have large cow size, produce more milk per day, sale more milk raw and earn higher income at household level. However, these households also have a higher dairy workload that has disproportionally gone to the women. Moreover, intra-household milk income share studies show that milk income control has moved from women to men compared to the experience in non-adopter households.

Time is a critical resource in any production activities (Ilahi, 2000). The household is an institution of social control that governs time allocations of its members between different activities (Fairhead and Leach, 2005; Fukuda et al., 2005). Time devoted to one activity competes with time devoted to alternative activities (Skoufias, 1994). Fafchamps (2001) argued productive household members divide their time between labor for own income-generating activities, labor provided to others and leisure, and they are willing to provide labor only if they receive a compensation that is at least equal to what could be earned in the off-farm labor market. The labor supplied by household members should therefore be proportional to the amount of benefit generated from this activity.

Therefore, it is assumed that those who invest more time/labor in a production system and work hard will produce more and control better resources that can change the course of their lives, i.e., they attend the end goal of ending poverty faster than those who invest less time in this system. The result of this study however shows otherwise. It shows that women who invest more time in dairying lost share of their milk income in dairy technology adopter households. On the other side, men invest more time in dairying in technology adopting households but also earn significantly more share of the dairy income compared to both men in non-adopter households and the time they invested in the production activities. The increasing women's workload is believed to be associated with shifting cattle management systems. In intensive dairying with crossbreed cows and new cut-and-carry fodder technology, livestock management has shifted from free grazing to stall feeding the cows in the backyard, an activity that falls within women's domestic chores.

It is also clear from this study that dairy technology adoption involves defeminization of the dairy income control. Dairy technology adoption implies potential reduction in women's dairy income share and, hence, income under the control of the women in relative terms. Women in technology adopter households invest significantly more time in dairying compared to women in non-adopter households. However, their dairy income share is declining. The increasing dairy income disproportionately goes to men. This does not only lead to intra-household income inequality between men and women but also has negative consequence for the welfare of other household members, and it challenges the assumption that development programs and policies that increase income at household level benefit all household members. Previous research (Adato et al., 2000; Mahmud, Shah and Becker, 2012) suggests that women's income share in household income determines their participation in household decision making and their intra-household bargaining power. Moreover, Agarwal (1997), Maluccio and Quisumbing (2000) and Njuki et al. (2011) indicated that income in the hands of women enhances the welfare of household members more than income in the hands of men. Hence, the income shift from women to men with dairy technology adoption in Ethiopia can negatively affect household members' welfare.

In conclusion, the finding of this study suggests that dairy technology adoption is associated with defeminization of milk income while it increases women's workload at household level. Hence, improving women's access to productive resources and technology information, strengthening women's participation in the formal milk market, ensuring women's control over the dairy income and introducing technologies that can reduce their workload can lessen the defeminizing effect of dairy technology adoption in Ethiopia and improve the socioeconomic status of women as well as the welfare of their household members.

Acknowledgments

This paper is part of the Ph.D. research project "Gender, intra-household dynamics and smallholder milk market participation in Oromia, Ethiopia," conducted in Selale area between 2013 and 2015. I gratefully acknowledge the financial support for this study from the Netherlands Organization for International Cooperation in Higher Education (Nuffic) and the International Foundation for Science (IFS). I appreciate the support I received from Ambo University (Ethiopian) and Radboud University Nijmegen (the Netherlands) during my Ph.D. study. I also thank individuals and organizations who generously provided the necessary information for this study.

References

Adato, M., Bénédicte de la, B., Mindek, D. and Quisumbing, A. (2000). "Final Report: The Impact of PROGRESA on Women's Statusand Intra-household Relations" International Food Policy Research Institute. Washington DC, USA.

Agarwal, B. (1997). "Bargaining" and gender relations: Within and beyond the household. *Feminist Economics*, 3(1), pp. 1–51. doi:10.1080/135457097338799

Austin, P.C. (2011). An introduction to propensity score methods for reducing the effects of confounding in observational studies. *Multivariate Behavioral Research*, 46(3), pp. 399–424. doi:https://doi.org/10.1080/00273171.2011.568786.

Caliendo, M. and Kopeinig, S. (2005). *Some practical guidance for the iimplementation of propensity score matching: Discussion Paper Series 1588*. Bon, Germany: The Institute for the Study of Labor (IZA).

Caraway, T.L. (2006). Gendered paths of industrialization: A cross-regional comparative analysis. *Studies in Comparative International Development*, 41(1), pp. 26–52. doi:https://doi.org/10.1007/BF02686306.

Chagwiza, C. (2014). *Engaging cooperative farmers in agricultural intensification: Case studies on honey, dairy and linseed value chains in Ethiopia. sfer.asso.fr*. Doctoral Thesis, Radboud University Nijmegen, The Netherlands.

Coles, C. and J. Mitchell 2011: Gender and agricultural value chains – a review of current knowledge and practice and their policy implications. ESA Working Paper No. 11-05. Online available at: http://www.fao.org/docrep/013/ am310e/am310e00.pdf, 07/12/2013.

CSA. (2008). *Population and housing census of Ethiopia 2007*. Preliminary Report. Addis Ababa, Ethiopia: Ethiopian Central Statistical Agency.

De Hoop, T. J. (2012). *How context matters for development effectiveness: A study into social norms and heterogeneous impacts* (Unpublished Doctoral Thesis). Radboud University Nijmegen, The Netherlands.

Dito, B. (2011). *Essays on women's bargaining power and intra-household resource allocation in rural Ethiopia* (Unpublished Doctoral Thesis). Erasmus University of Rotterdam, Institute of Social Studies (ISS), The Netherlands.

Doss, C.R. (2001). Designing agricultural technology for African women farmers: Lessons from 25 years of experience. *World Development*, 29, pp. 2075–2092.

EADD. (2009). *Strategy for integrating gender in EADD*. East Africa Dairy Development Project, Nairobi, Kenya, 1–11. Available at: https://cgspace.cgiar.org/bitstream/handle/10568/1965/Gender Strategy-East Africa Dairy Development Project.pdf?sequence=2.

Endeley, J. (2001). Conceptualising women's empowerment in societies in Cameroon: How does money fit in? *Gender & Development*, 9(1), pp. 34–41.

Fafchamps, M. (2001). Intra-household access to land and sources of inefficiency: Theory and concepts. In: A. de Janvry, G. Gordillo, E. Sadoulet, and J. P. Platteau, eds., *Access to land, rural poverty, and public action*, pp. 68–96. Oxford: Oxford University Press.

Fairhead, J. and Leach, M. (2005). The centrality of the social in African farming. *IDS (Institute of Development Studies) Bulletin*, 36(2), pp. 86–90. Available at: http://onlinelibrary.wiley.com/doi/10.1111/j.1759-5436.2005.tb00202.x/abstract.

FAO. (2010). *Investing in skills for socio-economic empowerment of rural women*, pp. 14–16. Gender and Rural Employment Policy Brief no.2. Gender Pathways Out of Poverty- Rural Employment, Rome, Italy. Available at: www.fao.org/docrep/013/i2008e/i2008e02.pdf [Accessed 20 October 1988].

Feleke, G. (2003). *Milk and dairy products, post-harvest losses and food safety in Sub-Saharan Africa and the Near East*. A Review of the Small Scale Dairy Sector – Ethiopia. FAO Prevention of Food Losses Programme. Rome, Italy: FAO.

Fukuda, D., Nepal, K.P. and Yai, T. (2005). Microeconomic models of intra-household activity time allocation. *Journal of the Eastern Asia Society for Transportation Studies*, 6, pp. 1637–1650.

Heinrich, C., Maffioli, A. and Vázquez, G. (2010). *A primer for applying propensity – score matching*. Impact Evaluation Guidelines. Strategy Development Division, Technical Notes No. IDB-Tn-161. Washington, DC. Inter-American Development Bank.

Ilahi, N. (2000). *The intra-household allocation of time and tasks: What have we learnt from the empirical literature ? Policy research report on gender and development* (No. 13). *Gender and development*. Working Paper Series No. 13, World Bank. Available at: http: //www.worldbank.org/gender/prr.

Kucera, D. and Tejani, S. (2014). Feminization, defeminization, and structural change in manufacturing. *World Development*, 64(August), pp. 569–582. doi:https://doi.org/10.1016/j.worlddev.2014.06.033.

Lenjiso, B.M. (2016). *Gender, intra-household dynamics and smallholder milk market participation in Oromia, Ethiopia*. Doctoral Thesis, Radboud University Nijmegen, The Netherlands. Available at: http://hdl.handle.net/2066/157487.

Lenjiso, B.M., Smits, J. and Ruben, R. (2015). Smallholder milk market participation and intra-household time allocation in Ethiopia. *European Journal of Development Research*, 27(5). doi: http://dx.doi.org/10.1057/ejdr.2015.54.

Mahelet, G.F. (2007). *Factors affecting commercialization of smallholder farmers in Ethiopia: The case of North Omo Zone, SNNP region*. Paper presented at the Fifth International Conference on the Ethiopian Economy, Addis Ababa, June 7–9, 2007, (May).

Mahmud, S., Shah, N. and Becker, S. (2012). Measurement of women's empowerment in rural Bangladesh. *World Development*, 40(3), pp. 610–619.

Maluccio, J. and Quisumbing, A. (2000). *Intrahousehold allocation and gender relations: New empirical evidence from four developing countries*. FCND Discussion Paper. International Food Policy Research Institute. Washington DC, USA.

MoFED. (2010). *Growth and transformation plan, 2010/11–2014/15, Volume I: Main text*. Ministry of Finance and Economic Development, FDRE, Addis Ababa.

Mutinda, G. (2011). *Stepping out in the right direction: Integrating gender in EADD*. A presentation prepared for the Gender and Market Oriented Agriculture (AgriGender 2011) Workshop, Addis Ababa, Ethiopia, 31st January-2nd February 2011. Nairobi (Kenya): East Africa Dairy D, (February).

Njuki, J., Kaaria, S., Chamunorwa, A. and Chiuri, W. (2011). Linking smallholder farmers to markets, gender and intra-household dynamics: Does the choice of commodity matter? *European Journal of Development Research*, 23(3), pp. 426–443. doi:https://doi.org/10.1057/ejdr.2011.8.

Skoufias, E. (1994). Market wages, family composition and the time allocation of children in agricultural households. *The Journal of Development Studies*, 30(2), pp. 335–360.

Staal, S.J., Nin Pratt, A. and Jabbar, M. 2008. *Dairy development for the resource poor. Part 2: Kenya and Ethiopia. Dairy development case studies*. FAO/PPLPI Working Paper, No. 44-2. Rome, Italy: FAO, p. 52.

Tangka, F., Ouma, E.A., Staal, S.J. and Shapiro, B. (1999). *Women and the sustainable development of market-oriented dairying: Evidence from the highlands of East Africa*. Paper presented at the International Sustainable Development Research Conference held at University of Leeds, Leeds, UK, 25–26 March 1999.

Timmer, C.P. (1997). Farmers and markets: The political economy of new paradigms. *American Journal of Agricultural Economics*, 79(2), pp. 621–627. doi:https://doi.org/10.2307/1244161.

Udry, C. (1996). Gender, agricultural production, and the theory of the household. *Journal of Political Economy*, 104(5), pp. 1010–1046. doi:https://doi.org/10.1086/262050.

9 Implementing "gender equity" in livestock interventions

Caught between patriarchy and paternalism?

Katie Tavenner and Todd A. Crane

In one of the first assignments for my gender research post-doctoral fellowship, I was tasked with traveling from Nairobi to rural Western Kenya to interview various key stakeholders on the existing barriers to achieving gender equity in the semi-intensive dairy sector.[1] After spending a few days meeting with project partners, I spent several weeks speaking with processing plant management, livestock extension officers and dairy farmers in the region. In one of my earliest facilitated group discussions, the issue of "fairness" in intra-household sharing of dairy income from formally marketed milk came up. What would a "fair share" look like? I asked.

> I don't want a 50/50 with my husband. If I get 20% of the [formally sold] milk income, that would be wonderful. I can buy small things for the house and the secret things my husband cannot buy for me. . . . Are you giving any trainings on chickens?
>
> (Middle-aged married woman, Aldai Mooi co-op discussion group, Nandi County, Western Kenya)

This was a response I did not expect. And yet the more I pressed on with why women and men in the discussion group would not find a 50/50 equal split desirable, the more I began to question an otherwise inherent assumption of my research – that there was a shared frame of reference of what "gender equity" is. Could "gender equity" in the dairy sector really be construed as wives receiving 2/10 of the total household milk income (and expanding their chicken production)?

For the remainder of my research in Western Kenya, I sought to answer these questions and explore how the meanings attached to "gender equity" could vary across different social groups and institutions within the Kenyan dairy sector. Moreover, I was interested in how these divergent perspectives were embedded in gendered power structures and whether and how these would influence the level of engagement and desirability for dairy development interventions.

The assumption that patriarchal agrarian societies need to be transformed to accommodate fairer and more balanced gender (and market) relationships is the moralistic lynchpin holding together the logic of gender equity development interventions, including those that have taken place in the Kenyan dairy sector. But what if "beneficiaries" do not define their "gender problem" in the same way as the one underpinning the above rationale (e.g., that local culture framed as "patriarchal" and "traditional" are inherently problematic)? What are the implications for development programs and practitioners aimed at delivering "gender-equitable" interventions if the underlying logic for problematizing these relations in the first place is not shared? Finally, if we recognize that there are differing visions of gender equity, how can development interventions ensure that the diversity of perceptions will be considered in implementation as opposed to relying on simplistic and exogenous notions of equity?

These are questions that I grappled with during my post-doctoral research and continue to wrestle with as a development practitioner. It is my hope that this chapter will open a dialogue around uncomfortable questions and areas of tension surrounding the implementation of "gender equity" in development interventions.

Introduction

Achieving gender equity in livestock value chains is lauded as an international development goal in global policy circles (Kantor et al., 2015; Pyburn and van Eeerdewuk, 2016). While recent studies (Chanamuto and Hall, 2015; Leder et al., 2017) have highlighted the complexity of women's empowerment objectives in agricultural and natural resource interventions, the underpinning logic of "gender equity" remains relatively unchallenged. The concept of gender equity – defined here as the ability for men and women to fairly participate in and derive benefit from market participation – has increasingly been seen as a cornerstone to successful value chain development (KIT et al., 2012; Lee et al., 2015; Sririam, 2018). However, the goal of gender equity is complicated in many livestock interventions because of the socially embedded gender hierarchies and practices that ascribe male privilege to certain livestock species, commodities and commercial livestock enterprise (Tavenner and Crane, 2018).

In Kenya, critical aspects of dairy production practices including cattle ownership, decision making over production and control of dairy income are commonly dominated by elder men (Baltenweck et al., 2016). While women and young men play a significant role in contributing labor to daily dairy tasks, the type of material benefits they derive often depends on the "man in charge" (husband, father, and/or other male relative). This is particularly the case as dairy production becomes commercialized through formal milk market sales (Tavenner et al., 2018). Recent studies (Farnworth, 2015; Njuki et al., 2015) have suggested women and youth may access benefits of intensification less directly (e.g., enhanced food security from higher milk yields, increased total household income, greater investment in education), even if they do not receive commensurate income from their labor. Aside from these potential benefits, the fact remains that increasing milk production through dairy intensification will impose additional labor burdens on Kenyan women.

To mitigate these potentially negative aspects in intensifying production, development interventions are pursuing ways to increase gender equity in the dairy sector. Gender equity in the Kenyan dairy sector has taken shape in livestock interventions as encompassing at least three key areas: (1) Improving women's access to assets and markets, (2) increasing women's participation in leadership and decision making and (3) improving women's share of dairy proceeds and balancing the division of dairy labor.

This chapter investigates how localized gender relations in the Kenyan dairy sector are identified as "problematic" by development practitioners, dairy management and farmers and the related "solutions" that have been implemented in the name of diminishing these "gender equity problems." The dynamics between local gender norms and dynamics and how these relationships intersect with development interventions lie at the heart of my critical feminist reflection and questioning of what it means to be "caught between patriarchy and paternalism." This first piece of analysis involves an interrogation of "patriarchy" – used here to refer to a development narrative that views "gender problems" through a normative lens whereby women are subjugated under the authority of men in an oppressive system of gender disparity (Mahler et al., 2015).

Invoking the "problem of patriarchy" as rationale for development interventions generally looks at gender as being binary, unhistoricized and unchangingly monolithic to account for an all-encompassing gender oppression. Since the mid-1990s, the term "patriarchy" has been largely abandoned by gender studies scholars, replaced by "intersectionality" in terms of a conceptual lens for understanding gendered inequalities (Patil, 2013). However, development interventionists have tended to rely on a traditional notion of patriarchy as rationale for increasing the participation of women in livestock interventions and for closing the global gap in agricultural and livestock production (FAO, 2016). In development practice, the concept of a global patriarchal oppression is contested by localized masculinities and femininities, social processes that reject "patriarchy, either as an over-simplification which does not account for how women uphold, subvert, or challenge those norms, while many others, both men and women, reject the notion that patriarchy is itself problematic.

However, addressing the "problem of patriarchy" through development interventions quickly bumps up against the countervailing "problem of paternalism." "Paternalism" in development logic encapsulates the notion that development practitioners act on/assume development goals that do not necessarily align with the goals or interests of their intended "beneficiary target groups" (Bakewell, 2008). Paternalistic interventions fail to recognize the rationale for people's decisions, which are embedded in cultural and social meanings and institutionalized codes of conduct. In addressing the "problem of patriarchy," paternalist logic usually devises market-driven solutions that view gender equity as an investment, whose benefits can be measured in terms of economic efficiency and improved development outcomes (Roberts, 2015). The result of this line of thought has been simplified, narrow and instrumentalist solutions to complex social issues. "Paternalism" is used in this chapter to refer to the material disconnect between development intervention goals and locally held cultural beliefs, social norms and institutions.

This chapter explores this push–pull dynamic of being "caught between patriarchy and paternalism" using critical feminist analysis and reflection drawn from a qualitative systems analysis of the Kenyan dairy sector. Data is drawn from key informant interviews with corporate officials, processing plant managers, dairy extension officers and co-operative members, and men and women dairy farmers in Nandi and Bomet counties of western Kenya.

Methods

The study used a systems-level qualitative research design to canvass the perspectives of different dairy stakeholders, including dairy farmers, milk processing plant managers, farmer co-operative leadership, dairy extension officers, local government officials and NGO staff in Nandi and Bomet counties in Kenya's North Rift Valley. Key informant interviews and facilitated discussion groups took place during two field visits from April–May 2016 and September 2016. The 34 key informants (16 women, 18 men) were identified through purposive sampling after initial meetings with local government officers. Interviews were semistructured and covered the array of gender issues the organization or individuals were involved/interested in.

A total of 22 facilitated discussion groups (7 women-only, 2 men-only, and 13 mixed groups) were held with dairy farmers and co-op leadership boards in Nandi and Bomet, which were organized through the assistance of NGO staff and by managers of individual cooling plants. The group talks centered on discussions of gendered participation, benefits and constraints in dairy production. For both individual and group interviews, the process of explaining the purpose of the research, potential risks and benefits and informed consent was done at place of interview and/or discussion group.

Linking critique and praxis is at the heart of feminist scholarship (Sultana, 2007; Ackerly and True, 2008). In this vein, the following offers a critical feminist reflection on power, agency and gender as a relational and intersectional concept in the development practice of implementing "gender equity." To bridge critique and praxis, I combine reflexive storytelling and qualitative research empirics from my experiences as a postdoctoral fellow in gender research working in the Kenyan dairy sector. The chapter is organized by introducing three domains in which gender equity interventions are being implemented – market access, participation and intra-household benefits sharing. My analysis of "patriarchy" in each of these areas involves an introduction to the localized gender norms and power dynamics that shape women's and men's placement in each of these domains, examining how the "gender problem" is framed by development practitioners, dairy management and farmers. The concept of "paternalism" is analyzed by presenting the types of "gender equity" development solutions currently being devised in the Kenyan dairy sector. The tension between the two is presented through critical feminist reflection and discussion of the disconnect between conflicting notions of the "gender problem" (patriarchy) and the material manifestation of "gender equity solutions" (paternalism) demonstrated in each of the three domains of dairy interventions.

Domain I: Market access

The first domain in which "gender equity" interventions are being pursued is in improving access to dairy assets and markets that have traditionally been exclusively for men. In key informant interviews and group discussions, there was a clear disconnect between how governmental agencies, dairy processors, producer organizations and farmers framed the issue of equitable access to dairy cattle. For example, in meetings with local government officers in Bomet and Nandi, the issue of cattle ownership was framed as something that was "a cultural issue in the past," but "nowadays anyone that wanted to own a cow, could." These male government officers, along with the male manager of the largest processing plant in Nandi, explained that the barrier to entry into cattle ownership was no longer cultural but rather dealt with the costs of purchasing and maintaining cows. This notion that cattle ownership exists outside of cultural norms was contradictory to the perspective I received in speaking with an all-women group of dairy farmers at a discussion group later that day.

> How can a wife own a cow? Impossible! Now single ladies, they're allowed to own them. They can just go and buy a cow. But a wife? Not in this community.
> (Middle-aged woman farmer, Kapkuto Village Dairy Group, Nandi)

I was struck by how these farmers invoked the concept of intersectionality – the notion that gender takes meaning from its intersection with other identities, roles and responsibilities – during our conversation. The women in the discussion made clear that the rules for engagement in dairy were quite different depending on a farmer's marital status. They highlighted that, while the issue of cattle ownership was generally not contested within households, decisions and benefit sharing around the sale of milk were highly contested. For example, they explained that in married households, husbands control the income derived from any milk marketed to a formal marketspace (usually in the form of farmer cooperatives), while wives were traditionally able to control "leftover milk," commonly used for home consumption or sold informally to neighbors and local shops. However, as local markets become more formalized and milk becomes a more profitable commodity, women's control of the "leftover milk" (how much to sell, to whom, prices and how to use the associated income) weakens significantly.

When I raised this issue of intra-household conflict around the sale of milk to dairy processor management, my questions were largely deflected, with one manager claiming that their mandate to ensure gender equity was already met, in that: "all of our cows are ladies!" (Middle-aged man, acting plant manager, NKCC, Nandi). The managers that did engage me in discussing gender issues mainly focused on their strategies for targeting women to become milk suppliers to the local cooperatives affiliated with the processing plants. In these cases, "gender equity" was invoked as a capitalist imperative, as "common business sense" and as a means to increase dairy processor profits by increasing the total number of milk suppliers.

Beyond the individual interviews with processing plant management, my experiences working in multiparty stakeholder meetings also revealed this pervasive reading of "gender equity."

Intervention I: The "business" case intervention

In October 2016, Unique Forestry convened a development stakeholder meeting with national-level government officials, Kenyan Dairy Board (KDB) management, research institutions (ILRI) and private-sector representatives to discuss how the drafting of the NAMA (Nationally Appropriate Mitigation Action Plan) could ensure the proposed draft was gender equitable and socially inclusive for smallholder dairy farmers.

Private-sector actors, supported by Unique Forestry and ILRI, were tasked with thinking of ways to make their businesses more "gender inclusive." Engaging with government stakeholders and private-sector players meant taking up the discourse of making the "business case" for gender equity. Emergent from this exercise was the pervasive bundling together of "women and youth" issues for inclusive business practices. A representative from Brookside, Kenya's largest private processing company, suggested giving youth motorcycles to transport milk deliveries to their processing plants. In this case, "youth" was code for "young men," since there are well-known cultural taboos associated with young women riding motorcycles (e.g., it's not "ladylike," dangerous, takes women literally out of the home). This paternalistic development solution of providing young men with transport to participate in the formal dairy value chain (to the profit of the processor in question) is an easier way to meet gender equity goals than fighting against/unpacking/de-legitimizing <u>men's power</u> in the dairy system.

This intervention example alongside the numerous interviews with processing plant managers revealed they are largely indifferent toward the issue of "gender equity," but some have taken tokenistic steps towards "women and youth inclusion." The overall premise that markets themselves are gender neutral and that men and women are already able to benefit from dairy production were common refrains amongst senior staff at processing plants. In most interviews, processor management claimed that ensuring women's and youth's inclusion in formal milk marketing was the responsibility of farmer cooperatives, not the processing plants. Their priority was simply reliable quantity and quality of milk delivered to their processing plants.

In sum, processor management adhere to the belief that everyone can participate in and benefit from the "gender-neutral" free market system, even when they are fully aware how cultural norms inhibit women's engagement in the sector. This appears to be a strategy of deliberate agnosticism of cultural gender norms through adherence to neo-classical economic ideologies and discourses.

Domain II: Participation

Interviews with farmer co-operative leadership stressed their commitment to gender equity was being fulfilled by integrating women into milk cooperatives as members and strengthening women's leadership within their formal decision-making committees. However, co-operative staff and farmers noted that the stigma around women participating in public politics (which was viewed locally as being a corrupt and "dirty" business) meant women that were interested needed men's permission to run for office.

> It's difficult to get women to run for office, or even attend meetings! Most of their husbands oppose, they feel threatened, so women won't be available. Our men are too proud, ego is the problem. Men want to remain men. We fear so much – fear sending our wives to school, employing them, fear that their position will become head of the household.
>
> (Middle-aged man, leadership board, Lelchego co-operative, Nandi)

Women's participation in farmer cooperatives is low, and the meaning of their participation is mixed. While a few women spoke to their ability of being able to rise within the organizations, they admitted that their ability to be in meaningful management positions (chairman, executive committees) was limited. Instead most are cloistered on "gender and youth" committees, and are only responsible for serving tea, saying the opening and closing prayers during meetings, and cleaning up afterwards.

Intervention II: The "scheduling" intervention

Both processing plants and farmer cooperatives sponsor field days a few times a year during which extension officers train farmers in production techniques, such as milk hygiene and improved cattle feeds. However, the approach taken by these two groups in scheduling these days differs. According to one processing plant manager, "We normally schedule trainings in the morning hours. I know the impacts this has on women's attendance, so the majority are men. But we get better [overall] participation in the morning." (Middle-aged man, NKCC processing plant, Nandi). This quote reveals that processor commitment to gender equity is effectively lip service, as they weigh total participation as more important than actively working to include those farmers most responsible for dairy labor.

Co-operative extension agents recognize women's labor contribution to dairying, and many try to specifically target women for trainings. However, in recounting past trainings, agents noted the difficulty in meeting their "gender equity" targets. "Mostly men attend the trainings. We are trying to encourage ladies to attend because when men attend, they don't adopt the trainings, and they leave women at home with no knowledge" (Young woman, community facilitator, Lessos co-operative, Nandi).

While these agents recognize that women are the primary dairy laborers, the notion that "gender equity" is can be achieved by a matter of scheduling is deflecting the fundamental gendered cultural dynamics around cattle.

Furthermore, while women are responsible for on-farm labor, and targeting the information to them could potentially lead to greater uptake of improved practices, many women lamented that when they do attend trainings and want to implement on farm changes, they lack the material resources and support of their spouse to change.

A limiting factor in women's participation in dairy trainings is the social norm that, "When a man is sharing equally with his wife, he doesn't feel like a man. If he's seen walking with his wife to the same meeting, he'll lose face. It is like his power has been taken" (Middle-aged man, Sirikwa dairy co-operative discussion group, Uasin Gishu).

Domain III: Intra-household benefit sharing

Interventions aimed at increasing women's benefit from engaging in dairy production in Kenya have tended to cluster around two areas. The first is through the creation of "gender-equitable" payment structures for milk sales. An example of this has been through signing women on as dairy cooperative members. As a member of the co-operative, you are given a bank account in which deposits are made bi-monthly or monthly by the co-operative. However, these payment structures can end up being tokenistic, as even when women have bank accounts in their name, they must be transparent about formal sales and report and oftentimes give over their earnings to their husbands. Thus, the direct payment method of the co-ops for increasing gender equity doesn't work for married women, as the distribution of income and control of assets are tied to masculine and feminine ideals around married women's and men's responsibilities.

The second strategy to stimulate changes in intra-household benefit sharing are commonly referred to as "gender sensitization" interventions. These interventions consider the potential ill effects of dairy intensification on women and bring farmers together to discuss intra-household allocation of milk income and the gendered division of labor.

Intervention III: The "sensitizing men" intervention

An international livestock-based NGO working in the region has staged a "gender sensitization" intervention to level perceived gender imbalances in intra-household labor and finance dynamics among smallholder dairy farmers. Using discussion-driven workshops, NGO staff known as "gender champions" use participatory activities to engage farmers, especially men, in discussing gender issues in dairy production. These workshops, also known as "gender sensitization trainings," were designed to address the perceived inequitable dynamics that emerge from women's engagement in marketing milk in the formal marketplace.

These trainings, delivered in two modules called "Gender and Family Focus" and "Sharing and Caring," use the logic of "different, yet complementary gender roles and responsibilities" to compel men to give their wives' a "fair share" of the formal milk income. The key message woven throughout these trainings is to bring attention to the cultural aspects of intra-household asset and income sharing in a non-threatening way. As one "gender champion" noted: "Our goal is to increase dairy production for everyone. Gender sensitization is required, not to force change, but to open people's minds" (Middle-aged woman, Gender and Social Inclusion Advisor, International NGO, Nandi).

The main activity used to "sensitize" male farmers was to have them complete participatory time clocks in small groups. During the activity, men would list all the activities they themselves and their wives perform over a 24-hour period. Since most of the daily reproductive labor, as well as the bulk of dairy labor (milking, feeding, watering, manure management) is done by women, men are prompted through the activity to "open their eyes to the sheer amount of time women spend on these activities, and to recognize that they are overloading women."

Using the logic of labor input versus financial returns as rationale for sharing and caring means men are essentially being paternalistically guilted into creating "more equitable" intra household sharing. On the one hand, this type of intervention reifies and reinforces the status quo and existing gender dynamics (traditional) that NGOs in the field have noted as being more palatable at the local level. However, the more radical notions of gender equality are lost.

There was a disconnect between paternalistic "gender sensitization" and Kalenjin people's perceptions of appropriate gender roles. As an intervention strategy, gender sensitization encompasses the notion that rural people aren't aware of what "gender" is. In reality, rural people are being sensitized to development practitioner notions of how gender relations "should be."

While "sensitization" was meant to convince men to include their wives in decision making on cattle sales, milk marketing and income expenditure, these proposed changes to locally embedded gender roles generally did not resonate with the men trained. The line of argumentation in the intervention is contradictory to the embedded gendered division of labor and patterns of household responsibility in the region – and crucially, the point that men feel entitled to allocated income as they see fit and the belief that men are already distributing income fairly in the household. "If I'm a husband and I get a dress for my wife, is she not benefitting from my increased milk income?" (Middle-aged man, County Livestock Officer, Bomet Government)

In addition to this common example in which affection and appreciation for women is shown through purchasing them clothes, women's "indirect benefits" from formal milk market participation have been used to argue that even if women aren't benefitting directly through controlled income of formally marketed milk, they can benefit in other ways. For example, men suggested if there is an increase in overall household income, and that money could be used to pay school fees or purchase more diverse food for the children, the household is benefitting through increased education and nutrition. However, it's problematic to assume that married women are benefitting indirectly if their husbands are the ones controlling milk income and if they disagree about spending priorities.

On the one hand, the "sensitization" intervention acknowledges that local masculinities exist and tries to engage with them in a non-threatening way to encourage benefit sharing through a more equitable division of labor in dairy households. On the other hand, the issue cannot be reduced to targeting men to recognize they are sharing unfairly in their households. Sharing income goes against what it means to be a man in the local context, and the prospect of loosening control of household incomes goes against meanings of headship, which are a crucial aspect of Kalenjin masculinity. A focus on trying to change the shares allocated or the amount of work women do ignores the embedded dynamic that dairy work is labor that falls into the category of "wifely duties" – activities that married women perform in enacting localized hegemonic femininity. A sole focus on labor also detracts from the central issue that any income derived from a commercial activity and especially one derived from cows (a masculinized asset) *is for men*.

Discussion

The interventions outlined in this chapter tended to ignore how localized constructions of masculinity around dairy production and household headship make a normative concept of gender equity undesirable, especially for married women. For example, dairy processors assumed individual decision making is desirable, but wives felt uncomfortable when their husbands weren't making the major decisions around dairy. The development assumption is that equality is practically achievable or obviously desirable, glossing over the cultural traditions that condition women's and men's desire to participate in dairy. The tensions between invoking "patriarchy" as rationale for development interventions and the resultant paternalistic "solutions" for "gender equity" described in this book chapter highlight three central themes for discussion. The first theme is that development interventions have largely dealt with focusing on "women" as a beneficiary group, as opposed to engaging in the structural, cultural and institutional ways in which gender is constructed. The second theme is that "gender equity" is not a normatively held concept, which has material implications for development solutions that use exogenous notions of equity to inform interventions. The final theme is that the bind of being "caught between patriarchy and paternalism" cannot be treated as merely a rural women's issue but a fundamental problematic embedded in gender development practice. Constant questioning of our values and underlying logic is central to the work of gender development practitioners – activities at the heart of feminist critique and praxis.

This study highlighted the pitfalls of focusing on "beneficiary groups" instead of engaging in the structural, cultural, and institutional ways in which gender is constructed and enacted in everyday life. Across all three interventions, the relational aspects of gender are rendered invisible in development and dairy management discourse, and instead homogenized groups of "women," "men" and "youth" are targeted. The reduction of "gender equity" to adding women and youth to dairy projects or to "sensitizing" men does not address the central issue of how the constructions of Kalenjin masculinity around cattle ownership and household headship precludes women's and youth's ability to participate in and benefit from dairy. Furthermore, this homogenization obscures the intersectional nature of gender difference, and the reality that dairy engagement and aspirations are patterned by how gender intersects with ethnicity, culture, marital status, age and wealth status. Indeed, most goals around gender equity target women instead of relational and intersectional gender power relations as the "problem." In effect, these approaches largely sidestep any tangible gender power issues. Moreover, the prioritization of engaging with social issues is well below the primary motivation of increased profit margins for dairy management, despite the attempts by development practitioners to push the idea of "making the business case for gender equity." These examples highlight that intervention strategies targeting women as a category of people to reach/empower are meaningless without recognizing gender foremost as a system of institutionalized social norms.

The second theme – that "gender equity" simply does not hold up as a universally held concept – has material implications for development "solutions" that use exogenous notions of equity to inform interventions and targets. The key logic underpinning these dairy interventions is that dairy farmers must be transformed to produce more milk, change complex social gender relationships within intervention timeframes and move homogenized "women" into positions of power within value chains they may have little to no interest in. This essential disconnect in visions of desirable intervention outcomes is key to understanding the "patriarchy" versus "paternalism" dynamic. The fact that farmers' sense of benefitting equitably from dairy production was not necessarily reflected as an equal share in dairy proceeds undermines the development logic of reducing objective gender inequality and highlights the socially constructed nature of "gender equity."

None of the three intervention strategies analyzed in this chapter addressed the gendered meaning imbued in dairy cows and the production of masculinized commodities nor the issue of wives' engagement in commercial enterprises as threatening to the local gender hegemony. As demonstrated in the study, no amount of "scheduling meetings at the right time of day" will solve, or even address, the fundamental cultural dynamics of patriarchy and cattle ownership. Throughout the study, Kalenjin farmers stressed that higher-priced commodities and commercial activities were meant "for men": "Cows, milk, tea, and maize – these things are for our husbands. Poultry and kitchen gardens are for wives" (Middle-aged woman, Lelchego co-operative member, Nandi). When questioned on why this was the case, farmers stressed that poultry and home vegetable production were easier for women to manage than dairy cattle and that women were able to retain control over the derivative income and resources from these activities, because "men consider chickens nothing, so they don't even ask about the profits!" (Middle-aged man, Cherobu co-operative member, Bomet). However, if production reaches "commercial levels" (the consensus in the counties defined "commercial" as keeping 200 chickens or more), women's control over proceeds would diminish, as this would constitute a "family enterprise." As articulated by one outspoken young female farmer:

> Women control all income from poultry and make all the decisions. Sometimes the income from poultry even beats dairy. More men will be approaching this business as it becomes more successful. Men have no clue how much women make [from chickens]. When they find out, they'll take over to commercialize!
>
> (Tinderet co-operative member, Bomet)

Thus, even when the seemingly "feminine" activity of poultry keeping is under the purview of women, there's a "grass ceiling" to their success. Gendered norms around commercialization as a masculine activity eclipse the norm of poultry production as a women's activity. In other words, while married women are encouraged to engage in these economic activities to help support their families, they are actively discouraged from trying to generate their own income that would contest the notion of household headship. It appears that in most cases, neither men nor women want that contestation. Because intensified dairy production is the combination of a livestock species and market process that are ascribed culturally masculine meaning, there is an inherent cultural conflict in reducing gender inequalities in these activities.

The final reflection point is related to the title of this chapter. The bind of being "caught between patriarchy and paternalism" cannot be treated as merely a rural women's issue. This dynamic is fundamentally embedded in gender development practice. In my experiences as a self-identified feminist from the Global North, the tension between respecting local cultural norms and the firm belief in gender equality is a source of constant questioning. Among development practitioners, gender equity is commonly seen as a stepping stone towards gender equality – however, this is problematic because it can reinforce paternalistic visions and interventions. This chapter has highlighted that while development practitioners often have specific notions of what constitutes equity, women and men smallholders do too. Indeed, gender-equitable arrangements are culturally specific, and the exogenous frames of equity based on ownership of assets, decision making, and related economic activities simply do not resonate in many parts of the world.

In my first year as a postdoctoral fellow, I received this advice from a dear colleague about a different approach to gender and livestock interventions: "The aim is not forcing change; the aim is to make people question when they want to go" (see Tavenner and Crane, 2016). While this perspective opens the gender and development agenda to move beyond problematic and paternalistic approaches to gender equity, there are trade-offs to this approach as well.

While many people working as gender and development practitioners are quite aware of these issues, operating within the programmatic logics associated with this field (process indicators, outcome targets, project deliverables, annual reports, monitoring and evaluation frameworks etc.) does not often create space for grappling with the tensions and contradictions of being caught between patriarchy and paternalism. It provides even less opportunity for strategizing how to find a path between them that avoids their pitfalls. If we actually start from the standpoint of rural people's own ideas about equity and their roles, lived experiences and agency in reconfiguring gender roles and resisting patriarchal norms, there's a chance that "they" will not move toward where we as development practitioners want to take them (whether it be sustainable intensification of livestock practices, participation in new value chains, or something else entirely). Development practitioners must engage with these tensions and questions and acknowledge that we are also in the process of being "caught between patriarchy and paternalism." This crucial exercise is at the heart of feminist critique and praxis, of reconciling internal conflict and moving toward an understanding that while interventions have gendered meaning attached to them, these meanings are also contested.

Conclusion

The assumption that gender relationships in agrarian societies are inherently problematic and must be transformed to accommodate "fairer" participation and benefit from neoliberal market relationships is the underlying rationale for many gender-equity development interventions, including those that have taken place in the Kenyan dairy sector. These interventions aimed at improving women's access to markets, assets and incomes have predominately targeted "women" as the "gender problem," and development "solutions" have largely failed to recognize the localized cultural systems of gender, livestock and market relationships that mediate farmer perceptions of what constitutes "gender equity." The combination of a fundamentally normative approach to gender relations with unequal power dynamics in interventions has enabled those in relative positions of power to promote "transformative" gender norms on others who do not share them.

Critical feminist praxis is an important exercise in reflecting on the resultant tensions and contradictions at the intersection of these power dynamics. While many development practitioners likely face structural constraints in programmatically addressing these uncomfortable questions, if we fail to engage in these difficult dialogues, we are at risk of missing the point of these interventions entirely. Thus, if there is to be a move away from paternalistic approaches to gender equity in development that are predicated on simplistic notions of patriarchal oppression, it may be better to start with considering what "gender equity" might look like for so-called "beneficiary groups," as well as to consider what institutional conditions in the sphere of development practice will enable more nuanced engagement with this difficult yet critical topic.

Note

1 Katie Tavenner and Todd A. Crane share co-authorship, but all first-person narrative references are by Katie Tavenner.

References

Ackerly, Bckerly B. and True, J. (2008). Reflexivity in practice: Power and ethics in feminist research on international relations. *International Studies Review*, 10, pp. 693–707.

Bakewell, O. (2008). "Keeping them in their place": The ambivalent relationship between development and migration in Africa. *Third World Quarterly*, 29, pp. 1341–1358.

Baltenweck, I., Omondi, I., Waithanji, E., Kinuthia, E. and Odhiambo, M. (2016). Dairy value chains in East Africa: Why so few women? In: R. Pyburn, ed., *A different kettle of fish? Gender integration in livestock and fish research*. Volendam: LM Publishers.

Chanamuto, N.J.C. and Hall, S.J.G. (2015). Gender equality, resilience to climate change, and the design of livestock projects for rural livelihoods. *Gender & Development*, 23, pp. 515–530.

FAO. (2016). *Developing gender-sensitive value chains – A guiding framework*. Rome: Food and Agriculture Organization.

Farnworth, C.R. (2015). *Recommendations for a gender dairy NAMA in Kenya prepared for lead and implementing partners to the Kenyan dairy NAMA*. Unpublished report 1–35.

Kantor, P., Morgan, M. and Choudhury, A. (2015). Amplifying outcomes by addressing inequality the role of gender-transformative approaches in agricultural research for development. *Gender, Technology and Development*, 19, pp. 292–319.

KIT, AGRI-PROFOCUS and IIRR. (2012). *Challenging chains to change: Gender equity in agricultural value chain development*. Amsterdam: KIT Publishers, Royal Tropical Institute.

Leder, S., Clement, F. and Karki, E. (2017). Reframing women's empowerment in water security programmes in Western Nepal. *Gender & Development*, 25, pp. 235–251.

Lee, J., Martin, A., Kristjanson, P. and Wollenberg, E. (2015). Implications on equity in agricultural carbon market projects: A gendered analysis of access, decision making, and outcomes. *Environment and Planning A*, 47, pp. 2080–2096.

Mahler, S.J., Chaudhuri, M. and Patil, V. (2015). Scaling intersectionality: Advancing feminist analysis of transnational families. *Sex Roles*, 73, pp. 100–112.

Njuki, J., Wyatt, A., Baltenweck, I., Yount, K., Null, C., Ramakrishnan, U., Webb Girard, A. and Sreenath, S. (2015). An exploratory study of dairying intensification, women's decision making, and time use and implications for child nutrition in Kenya. *European Journal of Development Research*, pp. 1–19.

Patil, V. (2013). From patriarchy to intersectionality: A transnational feminist assessment of how far we've really come. *Signs: Journal of Women in Culture and Society*, 38, pp. 847–867.

Pyburn, R. and Van Eeerdewuk, A. (eds.) (2016). *A different kettle of fish? Gender integration in livestock and fish research*. Volendam: LM Publishers.

Roberts, A. (2015). The political economy of "transnational business feminism." *International Feminist Journal of Politics*, 17, pp. 209–231.

Sririam, V. (2018). *Achieving gender equality and women's empowerment in smallholder adaptation: Lessons from IFAD's adaptation in smallholder agriculture programme*. CCAFS InfoNote. Wageningen, the Netherlands: CGIAR Research Program on Climate Change, Agriculture and Food Security (CCAFS), pp. 1–27.

Sultana, F. (2007). Reflexivity, positionality and participatory ethics: Negotiating fieldwork dilemmas in international research. *ACME: An International Journal for Critical Geographies*, 6, pp. 374–385.

Tavenner, K. and Crane, T.A. (2018). Gender power in Kenyan dairy: Cows, commodities, and commercialization. *Agriculture and Human Values*, 35, pp. 701–715.

Tavenner, K. and Crane, T.A. (2016). *Guide to best practices for socially and gender-inclusive development in Kenyan intensive dairy sector*. ILRI project report. Nairobi, Kenya: ILRI.

Tavenner, K., Fraval, S., Omondi, I. and Crane, T.A. (2018). Gendered reporting of household dynamics in the Kenyan dairy sector: Trends and implications for low emissions dairy development. *Gender, Technology and Development*, 22, pp. 1–19.

10 Implications of agricultural innovation on gender norms

Gender approaches in aquatic agriculture in Bangladesh

Lemlem Aregu, Afrina Choudhury, Surendran Rajaratnam, Margreet van der Burg and Cynthia McDougall

Introduction

Innovations in (aquatic) agriculture and consideration of gender norms

Gender equality is widely recognized as instrumental to achieving the Sustainable Development Goals (SDGs) in agricultural development programming. This outlook is strongly grounded in development research that for decades has generated evidence of gender disparities, for instance in access to productive resources, new technologies, knowledge, skills, markets, networks and services (Malapit et al., 2017). It has urged many programs and projects to address the perpetuating gender constraints and reduce the existing gender resource gaps. Yet recent efforts have come under critique for often only focusing on women's empowerment (Hillenbrand et al., 2015) or tangible aspects of gender gaps (such as in income, expenditure profiles, ownership of assets, control and authority over decision making) without addressing the root causes of gender inequalities, in particular underlying gender norms (Narayan, 2005). Kantor (2013) stresses that such a narrow perspective would block the potential contribution of programs to the reduction of gender inequality and poverty and limit the enhancement of sustained and equitable improvements of women's and men's livelihoods. Others state that innovation initiatives will have limited results in reducing poverty if these do not effectively engage with gender norms, gender relations and/or the gendered influences of formal and informal structures. These even risk reinforcing harmful gender norms, thus perpetuating gender inequalities (Cornwall and Edwards, 2010; Kumar and Quisumbing, 2010; Okali, 2011, 2012).

Studies have shown that gender norms and relations restrict the innovation capacity of women in agriculture, including aquatic agricultural systems (e.g., Philippa et al., 2016). Such literature has also revealed that innovations have advanced women's economic situation, especially through increasing their income. However, there is limited evidence on whether or how such economic improvements narrow the existing gender inequalities and benefit gender norms. Gender norms in this study mean socially constructed ideals of how women and men should behave in a given society. Gender norms are also time, culture, place and context specific (Knight and Ensminger, 1998; Marcus and Harper, 2014).

Therefore, by examining six communities of the CGIAR Research Program on Aquatic Agricultural Systems (CRP-AAS) in Bangladesh, this chapter explores the converse question: in what ways does or can agricultural innovation programming such as the CRP-AAS affect gender norms, and how can such an analysis help advance our understanding and integration of gender approaches?

Adaption to the gender integration continuum towards gender transformation

With the focus on effects of innovation initiatives on gender norms as the foundation of structural gender inequalities, this chapter aims to advance the understanding of how various gender innovation approaches are related to gender change.

For this purpose, we first introduce an adapted version of the gender integration continuum (see Figure 10.1), a conceptual framework introduced by the Interagency Gender Working Group (IGWG) to guide projects to qualify and assess their approaches to gender integration (Caro, 2009; IGWG, 2017; IFAD, 2018). Similar to the original framework, we clearly distinguish and position gender change in an integration continuum. As in the original framework, a differentiation between two extreme approaches can be seen at the top: gender blind and gender aware. The original further distinguishes exploitative, accommodating and gender-transformative dynamics. Presumably, gender-aware programs may be characterized by their growing effect on gender equality, even normative and systemic gender change, but that is not explicitly addressed in the graph. It is rather implied that accommodating gender approaches would lead to gender-accommodating change and gender-transformative approaches to gender-transformative change.

In our adapted version, we aimed at identifying the differences and overlaps between the approaches. We adapted the original graphic by explicitly distinguishing and connecting various innovation approaches and inserting how gender approaches of increasing gender sensitivity can be related to stages of the gender change continuum. This way, it provides more clarity and nuance to qualify gender-blind and -aware approaches in relation to their outcomes and impacts.

Changing WorldFish approaches and the adapted gender integration continuum

Since the qualitative data for this chapter were collected within a large aquatic agricultural systems program (CRP-AAS) from early 2013 to the end of 2015, led by WorldFish, we first introduce the gender approaches taken by WorldFish programs.

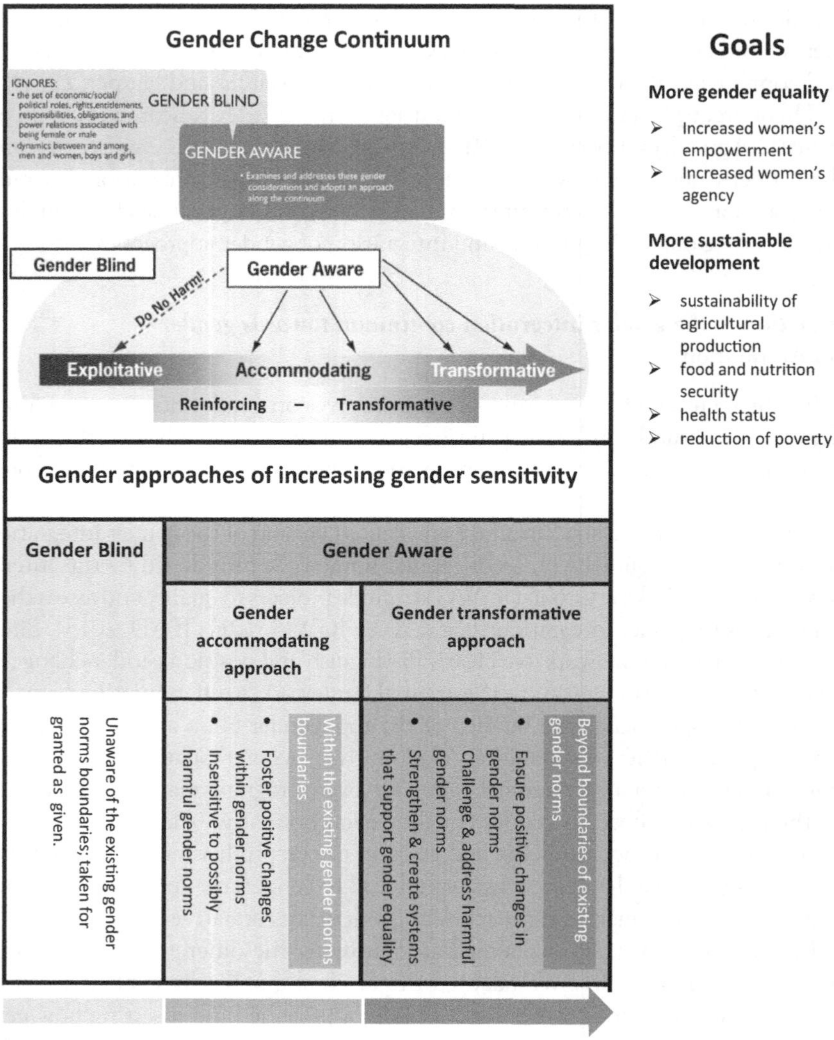

Figure 10.1 Gender Approaches and Change Continuum, applied for change in gender norms

Source: Adapted by the authors from the original Gender Equality Continuum Tool developed as part of the Gender Integration Continuum by the USAID Interagency Gender Working Group (IGWG) (e.g., Caro, 2009; IGWG, 2017).

WorldFish introduced innovations that are conducive to women in its programs. It has advanced innovation for women in especially vegetable cultivation and fish farming from the homestead, as these enterprises integrate well with prevailing cultural norms (Jahan et al., 2010). New technologies and management strategies were introduced to enhance women's production, income and nutritional benefits.

The CRP-AAS program, for instance, introduced an experiential learning approach for vegetable cultivation in a farmer-field-school-setting as participatory action research (PAR). It engaged participants as co-researchers aiming at empowering them and building ownership of the development process (Cole et al., 2014; Douthwaite et al., 2015). This process involved a close engagement of the front-line officers with the participants. They played a major role in convincing spouses and other family members to allow women's participation. The PAR activities allowed women and men to speak in public about their experiences, and their success helped in molding community attitudes around women's abilities (Apgar and Douthwaite, 2013; CGIAR AAS, 2013; Douthwaite et al., 2015).

Nevertheless, since the program did not make particular attempts to challenge existing gender norms, we label the CRP-AAS gender approach a gender-accommodating approach within the adapted Gender Integration Continuum Framework. Since then, WorldFish has also been using the Gender Integration Continuum (e.g., Lawless et al., 2017) and conducted studies to understand the gender impacts of its approaches. These studies have firmly established that there are harmful social norms and gender relations at play that restrict women in adopting innovations, even within what we would call an accommodative approach (Kantor et al., 2015; Morgan et al., 2015; Choudhury et al., 2017; Locke et al., 2017). However, there is still limited empirical investigation regarding whether or under which conditions a gender-accommodating approach can or cannot ignite positive impacts on shifting gender norms and gender inequalities, i.e., lead to gender-transformative change without any gender-transformative design (Farnworth et al., 2016; Choudhury et al., 2017).

Chapter organization

In the following sections, we first introduce the case study sites and approaches. This is followed by an analysis of the empirical data collected in 2014 in relation to the various approaches and changes in gender norms with help of the framework. It does so by presenting the gender norms that influenced men's and women's innovation and adoption, and reviewing the reported innovation outcomes in light of the gender accommodating approach used by the CRP-AAS. Then in the final section we will connect the results to what these can mean for further qualifying gender approaches in relation to gender-transformative change.

Bangladesh sites and study approaches from GENNOVATE

For this chapter, we use data of the CGIAR's GENNOVATE research initiative that was conducted by 11 CGIAR Research Programs (CRPs) in 26 countries. Its aim was to better understand how gender norms and agency influence men, women and youth of both genders to adopt innovations in agriculture and natural resource management (Badstue et al., 2014; Petesch et al., 2018).

The GENNOVATE data in Bangladesh were collected within the context of the CRP-AAS program in 2014 (Aregu et al., 2018b). Six villages in the Southwest Bangladesh polder zone (see Figure 10.2) were sampled from 16 communities

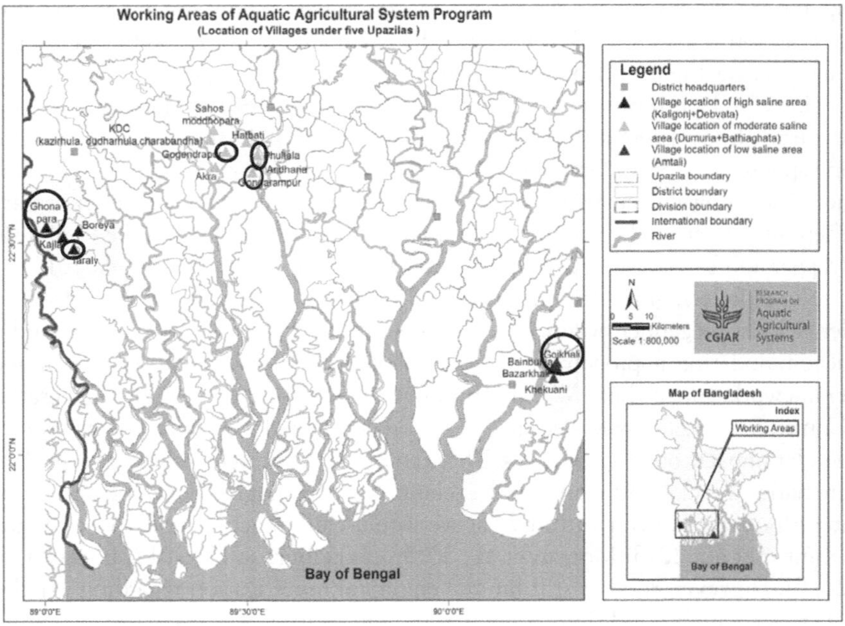

Figure 10.2 GENNOVATE study sites in Bangladesh. Circled: Satkhira (polder 3) – high-saline area; Khulna (Polder 29 and 30) – medium-saline area; and Barguna (Polder 42/F) – low saline.

Source: Aregu et al., 2018b, 15 and 16.

in four districts where WorldFish ran the CGIAR CRP-AAS in 2013–2015 (CGIAR-AAS, 2013; Cole et al., 2014).

The six selected villages represent the two major religions in the country and various levels of salinity, different sizes of population, economic dynamism, and gender gaps (Table 10.1).

A qualitative case study method was taken as a shared approach in the wider GENNOVATE research initiative. Its set up was more standardized than in a typical qualitative research approach to enable a comparative analysis across different CRPs. The methodological choices, data collection instruments and analysis are fully explained in Petesch et al. (2018). Coding and analysis of the qualitative data was done through NVivo 10 with a common coding structure.

A total of 36 focus group discussions (FDGs) were conducted with separate women's and men's groups (18 each). These were further differentiated by age/generation (adult male/adult female, and youth male/youth female), and the adults were again differentiated by wealth status (middle-income and poor). These were all guided by semi-structured questionnaires and tools. One tool is the 'Ladder of power and freedom' that sought to explore women's and men's

Table 10.1 Study sites and main characteristics

District	*Village#*	*Majority religion*	*Gender Gap*	*Economic Dynamics*	*Salinity*	*Population*
Khulna	1	Hindu	Low	High	Medium	Small
Khulna	2			Low		Medium
Khulna	5			High		Small
Barguna	4	Muslim	High	Low	Low	Medium
Satkhira	3				High	Large
Satkhira	6			High		

perceptions of their own decision-making power on a five-step ladder for the current time (both youth and adults) and for 10 years ago (only adults) (Aregu et al., 2018b; Petesch et al., 2018).

A total of 24 semi-structured interviews were conducted with female and male innovative farmers to gain in-depth insight in the types of innovations women and men engaged in, what enabled and hindered them in innovation adaption, and how they benefited.

Gender norms: innovation spheres, engagement and adoption

Gendered innovation spheres: around or away from the homestead

The analysis confirmed that spheres of innovation are strongly gendered in Bangladesh. When asked about the top innovations over the last five years, men and women identified different innovations as important. What women mentioned as enabling them to innovate ironically revolved around the normative constraints they face. For example, poor and middle-income women mainly valued innovations such as vegetable cultivation that were relatively easy to pursue within the homestead area, on a small piece of land, with little investment and aligned with their domestic duties. As enabling factors, they mentioned proximity and access to good soil and water quality, free resources like a river, as well as their ability to work hard. The women did not refer to financial capital, to which they have less access.

Men associated their most important innovations with rice cultivation, e.g., applying new varieties and machinery. This also has a normative foundation, since rice cultivation has been associated with men for many years and is perceived as a secure source to provide a livelihood for their families. Contrary to women, men reported as enabling factors those opportunities to which men usually get more access than women, like government extension and finances through credit as well as information.

Different among women and men was the importance given to innovations in other agricultural activities. Whereas all women mentioned innovation in poultry, livestock rearing and fish culture as important, only men from high-salinity villages – i.e., with high levels of salt in the soil – mentioned culture of high-value crustaceans, i.e., shrimp and prawn (see also Aregu et al., 2018a and 2018b).

Mobility constraints for women

Constraints in women's mobility came up as one of the key causes limiting their potential to innovate in aquatic agriculture. These are clearly related to the differences in spheres of innovations for women and men. As mentioned earlier, women favored innovations that can be carried out around the homestead, while men recalled success with innovations in a range of enterprises outside of the homestead.

When looking at sex-, age- and wealth-group-disaggregated variations across villages we see that in three villages, the female youth claimed that young girls experienced more freedom before marriage within the context of training and studies. One Hindu girl (village 1) explained that after marriage the in-laws would object. In contrast, male youth mentioned that young women cannot move freely, and in two villages they expressed that married women have more mobility. One young Hindu man (village 2) explained that no one would harass someone else's wife. Participants of all adult and youth FGDs though expressed their concern over young women's and girls' mobility when unaccompanied on their own, whether married or not. Also all groups from all six villages expressed that going to the market was problematic for all women in general.

The reasons behind women's mobility restrictions shed light on the deeper normative rationale of the communities that guided their practices. It was mentioned that husbands and family do not allow them to go to the market out of fear for women's safety (e.g., Eve teasing)[1] and respectability (talking to other men will hamper reputation). It was expressed as well that women's place was in the house and that there was no need to go out as long as able men were around. Some people expressed economic concern that women would be more easily tricked in bargaining for prices.

Nevertheless, participants from three village FGDs of the poor wealth status expressed that women no longer face as many barriers to their freedom of movement because financial problems let them care less about what others say, thus allowing women to go out of the household to earn income. Though, this might also indicate that barriers become looser in times of crisis with no guarantee for "better" times.

To be a "good wife" or a "good farmer"

Additionally, the analysis results highlighted that gender norms influence women's and men's opportunities and capability to be innovative farmers. For example, both women and men in the focus groups expressed that "good farmers" should be knowledgeable about new ways of farming and were therefore supposed to have good networks with information sources. Here, women fell short in opportunities for three interconnected reasons. As "good" wives, they were not allowed to freely move far from the homestead unless they were specifically called for trainings. Next, it was not acceptable that they would meet and talk with male strangers. Finally, they had to fulfill most of the family care roles. These all structurally limited them to fully benefit from what could widen their opportunities to innovate.

Gender-accommodating approach: positive outcomes with limitations

Promoting gender-inclusive interventions, as CRP-AAS did with vegetable PARs, proved to have a range of benefits. Below we first analyze the increasing scores of women in contrast to men on the "Ladder of Power and Freedom." Then we look into various benefits women have received as result of innovating.

Gendered steps on the Ladder of Power and Freedom

In general, the results of Ladder of Power and Freedom qualitative tool indicated that women's focus groups perceived that they had significantly stepped upwards on the Ladder of Power and Freedom over 10 years (Figure 10.3). In contrast, men stayed more or less on the same level over the same period.

Noteworthy is that change had not only different magnitudes but also varying start and end points (Figure 10.3). Women largely attributed this upward move from a very low level to their engagement in innovative agricultural practices. They expressed that earning more income made them gain more respect from their husbands. Access to knowledge and training were also reported as contributing factors. It could also be that they have gotten more power with age, but that was not mentioned.

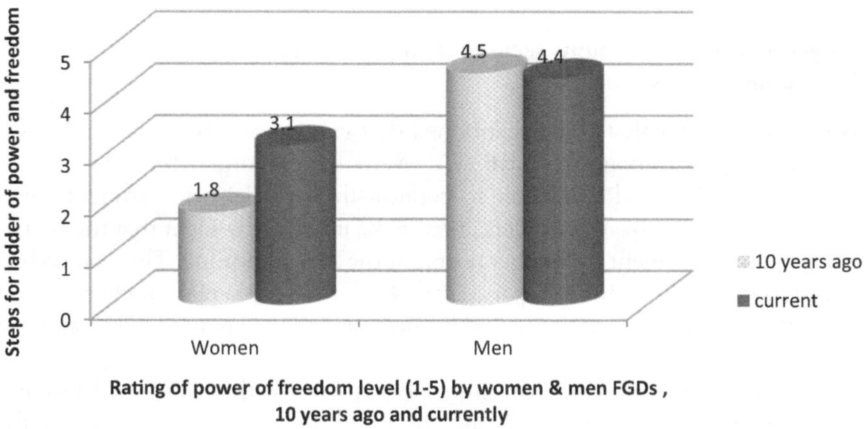

Figure 10.3 Ladder of Power and Freedom of women and men across all six villages as perceived in women and men's FGDs, 10 years ago and currently, levels 1 (lowest)–5 (highest)

Note: Step 1 or 2 stand for lacking or low power agency; 3 for moderate or growing power agency; 4 and 5 for rising or strong agency and power perceived).

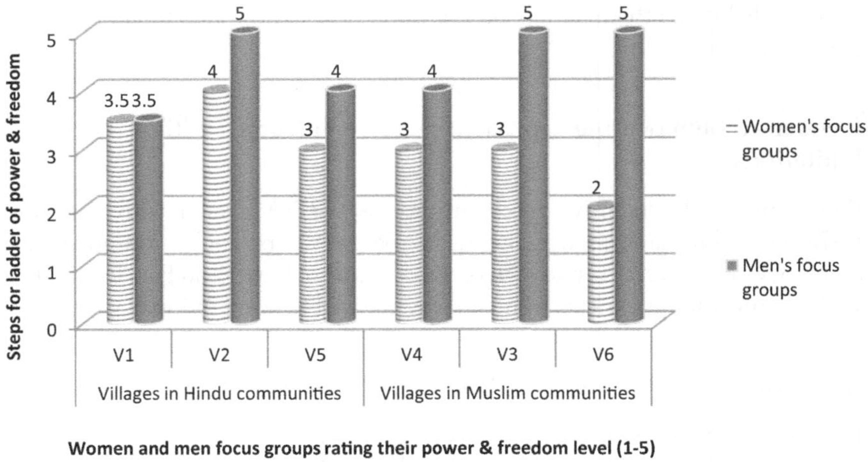

Figure 10.4 Ladder of Power and Freedom of women and men across all six villages as perceived in women and men's FGDs, levels 1 (lowest)–5 (highest)

Note: Step 1 or 2 stand for lacking or low power agency; 3 for moderate or growing power agency; 4 and 5 for rising or strong agency and power perceived).

Still, women rated their agency much lower than men. Only in Hindu village 1 did both women and men rate their agency at the same level (see Figure 10.4). In contrast, the gender difference of perceived power and freedom is extreme in Muslim village 6. More in-depth research is needed to further investigate possible explanations for the differences across the villages of the two religions.

Targeting women's domains secured their participation and innovation benefits

Both female and male farmers mentioned that gaining new skills, research tools and even inputs through the CRP-AAS were very important for their success. Women though found it difficult to confirm their participation straight away. From the female innovative farmers' interviews, it became evident that they experienced discouragement from other people at the very beginning. The women had to negotiate to get permission, particularly from their husbands, and first needed to work hard to prove that they could be successful producers and income earners:

> At first my husband did not agree. He was angry with me. When I was digging soil for gardening, he told me, "nothing will grow in that land. For nothing you are digging soil, nothing grows in the shaded land." He was laughing seeing my work (. . .); after Shumona[2] Apa [sister] had convinced him, he helped me. He helped me to dig the soil for gardening.
>
> (married female innovative farmer, 30 years, Hindu, village 1)

A male farmer portrayed a similar situation in his neighborhood. He told about a lady who wanted to grow ginger on her husband's land, as described below:

> The lady took out all of her husband's seed of eggplant from her homestead and borrowed ginger seed from me to plant. Her husband stopped her from doing that, and she told her husband that she wants to use this land for research (to try what she has been taught). For that reason, they had a huge quarrel. The husband came to me and returned all of my ginger seeds. But after some time, that lady convinced her husband and her son, and came again to take the seeds.
> (married male innovative farmer, 34 years, Hindu, village 1)

Increased women's income and family consumption

The first and most frequently mentioned benefit by the female innovative farmers was earning money from the sale of vegetable and fish. Some managed to circumvent their restricted market mobility by selling surplus produce to villagers from home. The women highly valued earning money that they used to meet daily household expenses:

> Men do earn always, now I got the opportunity to earn money by selling vegetables. Some cash is coming into my own hands. I am able to meet my son's needs. This is most important to me.
> (married female innovative farmer, 27 years, Hindu, village 2)

Also, the increase in production of vegetables helped women reduce household food expenditure:

> I am getting good and more vegetables. Now we are eating more vegetables and do not need to buy from the market. Most important thing is we don't need to buy vegetables from market with money.
> (married female innovative farmer, 28 years, Muslim, village 6)

Having more money was also seen as life changing. One female innovator told that it had reduced her dependence on others (married Muslim woman, 38 years, village 3). Another Hindu woman innovator explained that her husband became more co-operative in supporting her innovation because he realized that it brings in money (married, 44 years, village 2).

The women also told that the innovations enabled them to provide healthy meals as part of their tasks as "good" wives. Bangladesh women and girls often suffer from food insecurity and malnutrition (Smith et al., 2003). Having home-produced vegetables and fish regularly for family consumption was mentioned as a key benefit:

> (. . .) earlier we used to have fish once in a week. Presently we are having fish daily, our health becomes good by getting more nutrition (. . .) The

> important change we observed is our health has become good, we are getting nutrition from fresh vegetables daily (. . .)
>
> (married female innovative farmer, 30 years, Hindu, village 1)

Changes in women's confidence and gender relations

The improved economic situation also built women's self-confidence and respect. This also enhanced a local recognition of women as potential financial providers as opposed to the norms for "good" wives to be primary caregivers only. Their relations with their family members changed accordingly:

> the respect of the family for me has increased. For now, my opinion is also important and valued greater than before (. . .) Most important, the relation with my husband improved as my financial situation is improved. My husband loves me so much.
>
> (married female innovative farmer, 46 years, Muslim, village 3)

Illustrative is that a Hindu female farmer also mentioned that the relation with her husband improved after she had stopped asking for money for purchasing vegetables and fish for the family (married, 30 years, village 1).

Another woman narrated how the CRP-AAS program helped build her confidence in speaking in public:

> I was not able to talk to people properly. This year in the Maath Dibosh [Field Day] of WorldFish I addressed around 200 people.
>
> (married female innovative farmer, 27 years, Hindu, village 2)

Another female innovative farmer from village 1 mentioned that her older brother and some of her male neighbors copied her new ways of growing vegetables after noticing how she had benefited. A male innovative farmer of another village explained how his vegetable gardening was inspired by women in his neighborhood (single male innovative farmer, 26 years, Muslim, village 2).

Changing women's mobility constraints around training and knowledge sharing

Women's engagement in vegetable production and fish culture innovation also opened a window of opportunity to go out for trainings and meetings:

> I am earning some money; that is why my husband allows me now to go anywhere for meetings. He does not stop me if I want to do something (. . .) I can go anywhere if needed as I am earning some money nowadays.
>
> (married female innovative farmer, 25 years, Muslim, village 5)

The interviews revealed that the success of getting women participating in trainings and meetings helped to bend gender norms regarding women's mobility.

They were able to go out of the homestead to attend trainings. Female farmers recognized that their attendance in trainings and meetings allowed them to interact with other people and get the chance for knowledge sharing. Also, a young woman farmer explained that the training provided by WorldFish had helped her to go outside of the home, contact people easily and share her knowledge with others (married, 28 years, Hindu, village 6).

As a result, women also got involved in other economic and social group activities. For instance, one woman farmer said she entered the government initiative "Ekti Baari Ekti Khamar" (one house, one farm) project and explained how participating in WorldFish helped her (married, 30 years, Hindu, village 1). Similarly, another woman farmer (married, 33 years, Muslim, village 4) explained that it was easier for her to become a member of Sushilan[3] and the president of a CBO (community-based organization) because of the exposure, skills and confidence she gained from working in the CRP-AAS program. Unfortunately, the differences did not come out clearly and need further research.

Reinforcing gender norms: women around homestead and with more workload

In contrast to the female farmers who were growing their vegetables on their backyard plots, men were growing their vegetables on the larger, more commercial farmland. This gave men a better opportunity to gain maximum benefits from the innovations:

> I used to plant my crops in unorganized ways but now I am planting it in a line. For that reason my workload has reduced, various waste has reduced, I have got huge profit.
>
> (married male innovative farmer, 55 years, Hindu, village 1)

The men applied their new vegetable cultivation knowledge for paddy and other crop cultivation in the main farmland. For example, a male farmer explained he now also planted paddy seeds in a line, made a bed for betel leaves by using tape and took drainage measures, all as learned from WorldFish (married, 55 years, Muslim, village 4).

Men also took up vegetable production as an opportunity to make use of their farmland more efficiently. Meanwhile, they utilized family members' labor when needed, thus also putting more workload on the shoulders of women. One Muslim male farmer explained that his innovations did not affect his daily work. He had much barren land to use, and if he had no time for this work, his wife and other family members would take care of it (married, 40 years, Muslim, village 6).

Accordingly, some women clearly expressed the burden of their increasing workload:

> (. . .) this new work of gardening required to do more work daily, I got some physical problems. I do not have enough strength in my body, I get head

> ache, my legs and hands hurt. It is getting worse day by day. If I take medicine I feel better.
>
> (married female innovative farmer, 35 years, Muslim, village 5)

Discussion: gender-transformative innovation through a gender-accommodative approach?

The study provided insight into how the normative setting within which farmers operated differed quite significantly for men and women. Women were able to benefit but only within the limited scope of their limited land, capital, time and mobility. Although the CRP-AAS PAR process equally involved men and women and taught them all new ways of gathering, deciding and cultivating, the program was set up as sex segregated. This implied an approach that accommodated to what the specific groups had access to and what existing norms allowed: innovating men on large commercial land and women on small and often shaded, less productive land around the home. This can readily keep women in lower economic positions than men and thus perpetuate the existing gender inequalities or even enlarge the differences. For example, men's returns will easily become disproportionately larger than women's because the men have larger sizes of land they cultivate, while they can claim time of women to meet additional labor need. This supports what Hillenbrand et al. (2015) state: when examining individual change only, one fails to acknowledge influences on and from formal or non-formal institutional structures (Narayan, 2005). Therefore, we now review the results in light of the three inseparable indicators for gender-transformative change, i.e., agency, gender relations and structure (e.g., Hillenbrand et al., 2015).

Change in agency

Concerning agency, we see that women were able to earn an income or save consumption costs by engaging in a livelihood option. We found that women were able to keep the produce and the income derived from it. The PAR approach also gave women self-confidence and respect from the family and community. Many of the women innovators inspired other women and even men in their communities to copy or support the women innovators. This all reduced their dependence and financial burdens on their husbands (see Aregu et al., 2018a).

However, women explained in the FGDs that it was the homestead area that enabled them to pursue their innovations while taking care of the household work and obeying the mobility restrictions at the same time. Most women succumbed and were confined to utilizing their homestead land, which was often shaded and thus not optimal for vegetable production. None of them mentioned that they could make large purchases, but they rather used the little income they received from small pieces of homestead land for everyday household expenses. Some women even got into difficulty when trying to negotiate the use of family farmland.

Indeed, women also gained more mobility. They reported having gained freedom to attend trainings, share knowledge and/or join new NGO or government projects. However, the market remained a largely male-dominated domain, and gender norms still hindered women from going there.

Change in gender relations

In terms of gender relations, women reported that their relationship with their husbands had improved, mostly because of their successful income and food contributions to the household. Their family members were supportive to their efforts, particularly when they were successful. Women linked men's support to their new income, but it has not become clear whether men would similarly show support if women would try to take up non-stereotyped livelihood options or engage in other activities that challenge existing norms. Also, there was not any evidence of shifts in unpaid care work responsibilities, which assumedly still remained "women's" work.

Since women faced difficulties in getting access to quality land for their innovations, we see that their access to this key asset lagged behind that of men. Meanwhile, the CRP-AAS program had been able to rather equitably provide access to knowledge and skills. Although some conflicts reported about women being stopped to engage with the CRP-AAS program, front-line extension workers mostly resolved this. PAR's segregated women's and men's group approach supported women's engagement and enabled them to build trust and support among each other. But still, in the end, trainings were accommodating to women's domains. Also, the men were "allowing" the women to attend more training and projects. This already implied that women could not decide by themselves. Even if this may be positive as a small step to enhance their agency, it indicated that men kept on being the primary decision makers over women's time.

Structural change

Finally, the CRP-AAS innovation process enabled people to start building a positive image of women in the community as socially acceptable economic providers. Farmer field days and women's ability to demonstrate success brought this about without shaking the normative foundation of their communities. The existing gender norms on livelihood opportunities and choices for men and women stayed intact. As a result, the income that men and women are able to earn also differed. Stereotyped notions of what a man or woman can and cannot do and restrictions in women's mobility kept on making it difficult for them to profit from a wide range of access livelihood options, markets, resources and services as easily as men can. While the female innovators received recognition for their productive work, their reproductive works remained unassisted. Community attitudes and acceptance of women's entitlements, non-stereotyped livelihood choices, freedom of movement, equal decision making and thus gender equality have not been fully addressed or achieved. Like many other programs, it also has reinforced the notion that women's place is in and around the home, where she can earn some money within homestead boundaries while also maintaining her specific care responsibilities.

Conclusion: gender-accommodating approach as entry point?

Gender-accommodating approaches can function as practical entry points to induce women's empowerment and gender-transformative change. It is difficult to involve women in innovations when they have mobility restrictions and limited access to and control over assets. Hence an accommodating approach that supports women in innovation around their traditional domains and that respects the social norms can be a helpful strategy to get them on board with innovation programs. However, targeting women around innovations in women's traditional domains alone possibly reinforces existing gender norms. This perpetuates or may even increase gender inequalities, as women in comparison to their husbands are disadvantaged when it comes to innovations that require important resources like land while their labor burden increases. Clear differences between Muslim and Hindu villages were not found but certainly need to be further investigated in follow-up in-depth research (Choudhury and McDougall, 2018).

This study suggests that targeting women around innovations in women's traditional domains needs to be harnessed with a gender-transformative approach that includes critical reflections on gender norms beyond accommodating them. Further studies can help look into entry points to strengthen and sustain transformative changes. For distinguishing various ways of gender change and gender approaches, the adapted version of the Gender Integration Continuum has been very helpful.

More in-depth research is needed in testing various gender approaches and identifying their effectiveness for not only better gender change outcomes at the individual level but also sustainable gender-transformative impacts. This also implies the urgency to further identify indicators that articulate gender-transformative changes toward gender equalities. These indicators should reflect less tangible dimensions of change and address influential non-formal structures and broader contexts (historically, socially and politically) (Narayan, 2005 as referred to in Hillenbrand et al., 2015) Hillenbrand et al. (2015) found a wide range of holistic indicators for a gender-transformative approach. These were aligned to three domains, i.e., agency, gender relations and structure. These can further open the way to design and monitor how to address the underlying root causes of gender inequality and help unlock the potential of poor and marginalized women and men to participate in and benefit from innovation processes at a larger scale.

Notes

1 A euphemism used in South Asia for sexual harassment or assault, referring to Eve.
2 Shumona is the name of one of the AAS project officers/extension workers.
3 Sushilan is an NGO operating in those areas and also a partner of CRP-AAS.

References

Apgar, M. and Douthwaite, B. (2013). *Participatory action research in the CGIAR research program on aquatic agricultural systems*. CGIAR Research Program on Aquatic Agricultural Systems, Penang, Malaysia. Program Brief: AAS-2013–27.

Aregu, L., Choudhury, A., Rajaratnam, S., Locke, C. and McDougall, C. (2018a). Gender norms and agricultural innovation: Insights from six villages in Bangladesh. *Journal of Sustainable Development*, 11(4), pp. 270–287.

Aregu, L., Farnworth, C.R., Choudhury, A., Rajaratnam, S. and McDougall, C. (2018b). *Gender and innovation processes in integrated fish agri-food systems in Bangladesh and the Philippines: Insights from the CGIAR research program FISH*. GENNOVATE program report on the CGIAR Research Program FISH. Penang, Malaysia: WorldFish.

Badstue, L., Kantor, P., Prain, G., Ashby, J. and Petesch, P. (2014). *Innovation and development through transformation of gender norms in agriculture and natural resource management: A global comparative research initiative*. Concept Note for GENNOVATE. Montpellier: CGIAR. Available at: www.gender.cgiar.org/network-focus/gennovate/study-rationale-and-objectives [Accessed 14 June 2017].

Caro, D. (2009). *A manual for integrating gender into reproductive health and HIV programs: From commitment to action*. 2nd ed. Washington, DC: United States Agency for International Development (USAID) and Interagency Gender Working Group (IGWG). Available at: www.culturalpractice.com/wp-content/downloads/4-2009-2.pdf [Accessed 13 March 2018].

CGIAR Research Program on Aquatic Agricultural Systems. (2013). *Learning from implementation of community selection in Zambia, Solomon Islands, and Bangladesh AAS Hubs*. Evaluation and Learning Series Paper AAS-2013–24, Penang, Malaysia.

Choudhury, A. and McDougall, C. (2018). *Gendered ownership of aquaculture resources: Insights from two* villages *in Bangladesh*. Penang, Malaysia: CGIAR Research Program on Fish Agri-Food Systems. Program Brief: FISH-2018–19.

Choudhury, A., McDougall, C., Rajaratnam, S. and Park, C.M.Y. (2017). *Women's empowerment in aquaculture: Two case studies from Bangladesh*. Rome, Italy: Food and Agriculture Organisation of the United Nation, and Penang, Malaysia: WorldFish.

Cole, S.M., Kantor, P., Sarapura, S. and Rajaratnam, S. (2014). *Gender-transformative approaches to address inequalities in food, nutrition and economic outcomes in aquatic agricultural systems*. Working Paper AAS-2014–42. Penang, Malaysia: CGIAR Research Program on Aquatic Agricultural Systems.

Cornwall, A. and Edwards, J. (2010). Introduction: Negotiating empowerment. *IDS Bulletin*, 41(2), pp. 1–9.

Douthwaite, B., Apgar, J.M., Schwarz, A., McDougall, C., Attwood, S., Senaratna Sellamuttu, S. and Clayton, T. (eds.) (2015). *Research in development: Learning from the CGIAR research program on aquatic agricultural systems*. Working Paper AAS-2015–16. Penang, Malaysia: CGIAR Research Program on Aquatic Agricultural Systems.

Farnworth, C.R., Baudron, F., Andersson, J.A., Misiko, M., Badstue, L. and Stirling, C.M. (2016). Gender and conservation agriculture in East and southern Africa: Towards a research agenda. *International Journal of Agricultural Sustainability*, 14(2), pp. 142–165.

Hillenbrand, E., Karim, N., Mohanraj, P. and Wu, D. (2015). *Measuring gender transformative change: A review of literature and promising practices*. CARE USA. Working Paper.

IFAD. (2018). *How to do Design of gender transformative smallholder agriculture adaptation programmes: Gender and climate change adaptation*. Rome, Italy. Available at: https://www.ifad.org/documents/38714170/40215365/How+to+do-+Design+of+gender+transforma tive+smallholder+agriculture+adaptation.pdf/c5f3c4ff-26a3-4ac6-a3ed-3b8dfc93af56.

Interagency Gender Working Group (IGWG). (2017). *Gender integration continuum 2017*. United States Agency for International Development. Available at: www.igwg.org/wp-content/uploads/2017/05/FG_GendrIntegrContinuum.pdf [Accessed 13 March 2018].

Jahan, K.M., Ahmed, M. and Belton, B. (2010). The impacts of aquaculture development on food security: Lessons from Bangladesh. *Aquaculture Research*, 41(4), pp. 481–495.

Kantor, P. (2013). *The contribution of gender transformative approaches to value chain research for development*. Research Brief CGIAR Livestock 2. Nairobi, Kenya: ILRI.

Kantor, P., Morgan, M. and Choudhury, A. (2015). Amplifying outcomes by addressing inequality: The role of gender-transformative approaches in agricultural research for development. *Gender, Technology and Development*, 19(3), pp. 292–319.

Knight, J. and Ensminger, J. (1998). Conflict over changing social norms: Bargaining, ideology, and enforcement. In: M.C. Brinton and V. Nee, eds., *The new institutionalism in sociology*. New York: Russell Sage Foundation, pp. 105–127.

Kumar, N. and Quisumbing, A. (2010). *Policy reform towards gender equality in Ethiopia: Little by little the egg begins to walk*. IFPRI. Available at: http://siteresources.worldbank.org/EXTARD/Resources/336681-1236436879081/5893311-1271205116054/QuisumbingKumar.pdf [Accessed 15 September 2017].

Lawless, S., Doyle, K., Cohen, P., Eriksson, H., Schwarz, A.M., Teioli, H., Vavekaramui, A., Wickham, E., Masu, R., Panda, R. and McDougall, C. (2017). *Considering gender: Practical guidance for rural development initiatives in Solomon Islands*. Program Brief 2017–22. Penang, Malaysia: WorldFish.

Locke, C., Muljono, P., McDougall, C. and Morgan, M. (2017). Innovation and gendered negotiations: Insights from six small-scale fishing communities. *Fish and Fisheries*, 18, pp. 943–957.

Malapit, H.J., Sproule, K. and Kovarik, C. (2017). Using cognitive interviewing to improve the women's empowerment in agriculture index survey instruments: Evidence from Bangladesh and Uganda. *Journal of Gender, Agriculture and Food Security*, 2(2), pp. 1–22.

Marcus, R. and Harper, C. (2014). *Gender justice and social norms – Processes of change for adolescent girls: Towards a conceptual framework 2*. Overseas Development Institute (ODI). Available at: www.odi.org/sites/odi. org. uk/files/odi-assets/publicationsopinion-files/8831.pdf [Accessed 24 April 2018].

Morgan, M., Choudhury, A., Braun, M., Beare, D., Benedict, J. and Kantor, P. (2015). *Understanding the gender dimensions of adopting climate-smart smallholder aquaculture innovations*. Working Paper AAS- 2015–08. Penang, Malaysia: CGIAR Research Program on Aquatic Agricultural Systems. Available at: http://pubs.iclarm.net/resource_centre/AAS-2015-08.pdf.

Narayan, D. (ed.) (2005). *Measuring empowerment: Cross-disciplinary perspectives*. Washington, DC: The World Bank.

Okali, C. (2012). *Gender analysis: Engaging with rural development and agricultural policy processes*. Future Agriculture Working Paper 026.

Okali, C. (2011). *Searching for new pathways towards achieving gender equity: Beyond Boserup and women's role in economic development*. Agricultural Development Economics Division. ESA Working Paper No 11–09. Rome, Italy: FAO.

Petesch, P., Badstue, L. and Prain, G. (2018). *Gender norms, agency, and innovation in agriculture and natural resource management: The GENNOVATE methodology*. Mexico, DF: CIMMYT. Available at: http://gender.cgiar.org/wp-content/uploads/2018/02/GENNOVATE-Methodology_Feb2018_FINAL.pdf [Accessed 4 March 2018].

Philippa, J.C., Lawless, S., Dyer, M., Morgan, M., Saeni, E., Teioli, H. and Kantor, P. (2016). Understanding adaptive capacity and capacity to innovate in social – ecological systems: Applying a gender lens. *Ambio*, 45(Suppl.3), pp. 309–321.

Smith, L., Ramakrishnan, U., Ndiaye, A., Haddad, L.J. and Martorell, R.S. (2003). *The importance of women's status for child nutrition in developing countries*. IFPRI. Research Report 131.

Part 5

Gender, labor and decision making

11 Permanently seasonal workers

Gendered labor relations and working conditions of asparagus agricultural workers in Ica, Perú

María del Rosario Castro Bernardini

Introduction

Despite the importance of previous studies that unveiled the negative impacts of women's participation in the non-traditional agricultural exports (NTAEs), there has been relatively scant attention devoted to examining the complex labor relations and working conditions that intersect in the seasonal and gendered production of asparagus. The characterization of workers' employment statuses as temporary, although true, may not be sufficient to describe the complex matrix of workers' labor relations and gendered participation in an industry that operates year-round. Furthermore, the public portrayal of these workers as either a sign of economic progress or labor exploitation does not thoroughly consider workers' labor conditions in the context of what the literature defines as a seasonal labor, workers' gendered narratives of job satisfaction and the reasons behind their adherence to the asparagus industry. Built upon previous studies analyzing Peruvian asparagus agricultural workers, this chapter's main objective is to examine how asparagus agricultural workers experience the labor organization and labor conditions at asparagus farms. Furthermore, the study shows how these experiences differ by the gender of the workers.

The modernization and feminization of agriculture

In the late 20th-century, scholars noted the onset of a transformative process occurring in the agricultural production and consumption of food. This transformation relates to the rise of modern agricultural systems, such as contract farming and agribusiness with an export orientation (Marañon, 1993; Little, 1994; Preibisch and Grez, 2010; Sachs and Alston, 2010; Patel-Campillo, 2011). Following these processes, the Latin American region has undergone an unprecedented rise of their NTAE industries in the last three decades. The development of the NTAEs in Latin America s changed the structure of agricultural production and employment from subsistence agriculture towards the increasing participation of wage laborers in industrial agriculture (Ballara and Parada, 2009; Soto Baquero and Kelin, 2012a, 2012b).

In the same line, the literature's analysis of the feminization of agriculture has primarily addressed the increase in female wage labor as result of the modernization of agriculture and the emergence of the NTAEs since the 1990s (Collin, 1995; Appendini, 2002; Deere, 2005; Allen and Sachs, 2007; Ballara and Parada, 2009; Sachs and Alston, 2010; Patel-Campillo, 2012; Radel et al., 2012). The women's rights agenda as well as the international development agenda[1] have raised the demand for paid labor with the understanding that income allows women's economic independence, therefore narrowing the employment gap between men and women. Despite the existence of positive gains from women's participation in agriculture, different voices argued that an analysis of women's participation in the agricultural labor market in Latin America is needed, since there are instances that show that females' new role in paid agricultural activities is not necessarily improving their livelihoods but instead are creating new contexts of dependency and labor vulnerability (Soto Baquero and Klein, 2012a, 2012b).

The Peruvian asparagus industry

Peru's asparagus production was introduced in the north of the Peruvian coast during the 1950s. Initially considered experimental, asparagus was not produced as a commercial crop at the start of its production. It was not until the late 1970s that asparagus production rose due to modernization and industrialization. By the end of the 1980s, Peru had experienced the so-called "asparagus boom" with its exportation of "white gold" (white canned asparagus) to the Global North (Valcárcel, 2003). The development and success of this NTAE led to the creation of agricultural complexes throughout the Peruvian coast, extensively in the north to the south-central region.[2] Asparagus represents one of the leading crops in Peru's NTAEs, although in the last decade, producers have been diversifying their production with the cultivation of other crops such as table grapes (Chacaltana and Yamada, 2009; Meade et al., 2010). Particularly, Peru is a top world exporter of the fresh green asparagus variety, which is mainly destined for the U.S. market (Nichols, 1990; Díaz Rios, 2007; Benson, 2008; Chacaltana and Yamada, 2009; Meade et al., 2010; FAOSTATS, 2017).

The development process of the Peruvian asparagus industry appears to be related to the worldwide shift in agricultural organization experienced in the late 20th century. Different features of the global transformation observed in the literature appear also in the Peruvian asparagus industry. Particularly, the role of the workforce is central in understanding Peru's industrialization of agriculture, which employs nearly 200,000 workers (Agrarian Commission of the Peruvian Congress in Ruiz-Bravo and Castro, 2012). Workers' demands for increased labor rights, the improvement of working conditions and occupational health in the NTAEs as well as the debate around the modification of the Law of Agrarian Promotion (Law 27360) has become part of the national debate and has reached the Peruvian Congress.

Trabajadores asalariados (agricultural wage workers)

Studies on NTAE in Peru in general examined the labor relations within this sector by focusing primarily on wage workers [trabajadores asalariados]. Marañón (1993) defined wage laborers in asparagus production as "temporary peasants without land or those that have land but still need to supplement their income" (p. 28). Similarly, for Moura (2010) the notion of agricultural wage workers relates to the existence of wages or salaries, which define the relation between workers and agro-industrial firms. Moura (2010) has argued that Peruvian agricultural wage workers find themselves impoverished because they fail to receive a salary that allows them to provide for their basic needs (p. 101). Furthermore, national regulations, such as Law 27360 mentioned above, allows the employment of agricultural labor under temporal contract with the aim to formalize agricultural labor contracts and to create incentives for producers to formal hire labor for short-term periods (Chacaltana and Yamada, 2009).

Gender, vulnerability and agency in agricultural labor

Additionally, the analysis of the asparagus industry brings into conversation the gendered division of labor and its impacts on the lives of agricultural wage workers. This is because one of the central features of this sector's employment is the high presence of female temporary agricultural workers (Ferm, 2008; Flores Mego, 2011). Contrary to corporate discourses that indicate the preference of females for their ability to handle asparagus, gender scholars and activists pointed out that the presence of female wage workers in the agricultural sector emerges from fragile social conditions and from an adaptability to temporary work statuses (Collins, 1995; FEPROMU-ICA, 2007; Ferm, 2008; Sachs and Alston, 2010; Flores Mego, 2011). The fragile social conditions that made this group of workers a more vulnerable workforce can be put in conversation with the notion of "disposability" used by Wright when examining the feminization of labor in Mexico's maquilas (1999). Despite the different nature of both industries, this notion could be useful when examining the value that female and male agricultural workers give to their own work when working in the asparagus industry of Peru. By understanding agricultural workers' own perspectives regarding their labor's value, this study seeks to incorporate a transnational feminist approach (Mohanty, 2003), which, while examining critically the fragile labor conditions of agricultural workers, seeks to reflect on the value, agency and possibilities of transformation that this group of workers brings to the asparagus industry.

Methodology and research approach

This research employs an extended case study in asparagus-producing towns located in one of the major asparagus-producing districts located in the Province of Ica, within Peru's Ica Region. The rationale behind the use of an extended case study, given its systemic approach, aims to encompass a wider scope to understand the everyday lives of agricultural laborers (Burawoy, 1991). As the grounded basis of this approach, this research targets towns revolving around asparagus production.

The selection of the extended case study follows a non-probabilistic purposive sampling design with the aim of achieving depth and significance in its findings (Singleton and Straits, 2010, p. 172). The focus on major asparagus-producing zones implies the areas will have a high concentration of asparagus workers. By considering the diversity of growers among asparagus producers, a focus on the scale of production will capture the different types of labor organizations and their correspondingly different impacts.

Following qualitative methodology, field research was conducted during the major asparagus harvest, from November 2015 to January 2016. This chapter reports on the results obtained from field research observations and 42 semi-structured interviews with female and male workers. The interviews took place primarily in the two asparagus-producing towns selected for the study. The first town featured a large presence of small (from 3 to 5 hectares) and medium asparagus farms owned and managed by smallholders. The second village featured a small number, but large presence of large agricultural farms (more than 80 hectares) owned and managed by agricultural and agro-export firms.

Labor organization and working conditions in Ica's asparagus production

Who are the asparagus agricultural workers?

Both female and male laborers were interviewed with the aim of observing gender differences when examining the labor and working dynamics in asparagus production. Table 11.1, which outlines the distribution of workers interviewed by gender, shows that the number of female workers interviewed exceeds the number of male workers interviewed by more than 10%. The greater number of women interviewed matches the secondary data that established that around 60% of the workers were women (Ferm, 2008; Flores Mego, 2011).

Regarding demographics, the average age of interviewed workers was 35 years old, with the total range spanning 21 to 51 years old. In relation to workers' families and their offspring, the research shows that 83.4% of workers reported to be married or cohabiting, while 16.6% of workers, mostly women, reported to be divorced or separated. In the same line, among married or cohabitating couples, a

Table 11.1 Gender of the workers

Gender	*Frequency*	*Percentage*
Male	18	42.9
Female	24	57.1
Total	42	100

small group of mostly women reported to be separated or in process of separation (7.41%) but still cohabitating with their partners, while others reported to be in a relationship with their partners although they maintain a long-distance relationship (4.76%). Concerning workers' children, almost all workers had children of the age of 12 or under, while 61.9% of workers reported having children under six years old. Workers interviewed had children with a range from one to five children, on average 2.5 children, with no observable differences among male and female workers.

In contrast with previous studies that stated that agricultural workers are predominantly migrant workers from the Andes region to Ica, the interviewed workers in this study permanently lived in Ica, with almost three-quarters of the sample being born in this region with Spanish as their first language. When looking at gender differences, male workers reported higher rates than females of being from the Andean region and, consistent with their region of origin, have a native language as their mother tongue. In this study, the workers' status as permanent residents in the area appears to be key to understand their access and permanency to the asparagus labor market. It is important to note that the working experience of this group of workers may not reflect the experiences of migrant workers, for instance, but can bring new reflections on a workforce that permanently reside in Ica and not only migrate to this region in the harvest time. In this sense, this study argues that due to the proximity to major centers of asparagus production in the region – and therefore nationally and globally – the workers included in this study are not simply employed in asparagus farms but that their residence, occupation, family situations and access to land are largely defined and revolve around the development of the asparagus agricultural complex in Ica, Perú.

Among other sociodemographic characteristics of this group of workers, previous studies of Peruvian agroindustry workers stated that this population has a higher level of education in comparison to the national average (Valcarcel, 2003). This study indicates that 50% of the sample reported to have completed high school, around 20% of the sample reported to have attended some time on higher education and finally 7% reported finishing their studies at higher education institutions such as colleges or technical schools. The research shows that workers who reported to be born in the Andean region and had a native language as their mother tongue appeared to have the lowest level of education in comparison with the rest of the sample.

Regarding their labor participation in asparagus farms, workers interviewed hold employment in asparagus farms with different scales of production. Research results show that 38.1% of the workers interviewed were employed by small asparagus farms, 16.7% by medium farms[3] and 40.5% by large farms. It is interesting to note, however, that 4.8% of the workers reported working on both small and large asparagus farms at the time the interviews were conducted. To achieve diversity, the selected interviewees not only worked in different scales of production but also on asparagus farms owned or managed by different producers. Among the sample of workers interviewed, workers were employed by 18 different producers.[4] The different scale of production as well as the large number of asparagus farms in which laborers interviewed worked allowed to identify patterns, as well as assess differences, in the labor organization and working conditions of asparagus production.

Seasonal production: equals temporary and informal employment?

The temporary characterization of the employment offered in this industry is tied to the seasonal nature of the production of asparagus. Although produced year-round, asparagus harvest occurs at the research sites of this study for at least four to six months of the year, depending on the farm and scale of production. For this reason, it is throughout the harvest season when the industry requires a greater number of laborers, and therefore there is less of a demand for labor in the off-season. As a result of the high fluctuation of labor between seasons of asparagus production, the interviewed laborers employment status is most of the time defined temporally. As can been seen in Table 11.2, at least four-fifths of the sample report to be temporary workers, working through temporary or informal contracts.[5] Almost all the female workers and two-thirds of the male workers reported to work under a temporary employment status.

As is shown in Tables 11.3 and 11.4, when examining the employment status and type of contracts by scale of production, among workers employed at large farms, the existence of permanent position and written contracts is higher in comparison to small farms.

This is also true when examining the employment status by the gender of worker. As Tables 11.5 and 11.6 show, among male workers the existence of more permanent positions and written contracts is also higher.

Table 11.2 Worker's employment status

Employment Status	*Male*	*Female*	*Total*
	Percentage	*Percentage*	*Percentage*
Temporary	66.67%	95.83%	83.33%
Permanent	33.33%	4.17%	16.66%
Total	18	24	42

Table 11.3 Worker's employment status by scale of production

Employment Status	*Small*	*Medium*	*Large*	*Total*
	Percentage	*Percentage*	*Percentage*	*Percentage*
Temporary	100%	85.71%	68.42%	83.33%
Permanent	0	14.29%	31.59%	16.66%
Total	16	7	19	42

Table 11.4 Worker's type of contract by scale of production

Type of Contract	*Small*	*Medium*	*Large*	*Total*
	Percentage	*Percentage*	*Percentage*	*Percentage*
Oral	100%	71.4%	6.66%	55.55%
Written	0	28.57%	93.33%	44.44%
Total	14	7	15	36

Table 11.5 Worker's employment status by gender

Employment Status	*Male*	*Female*	*Total*
	Frequency/Percentage	*Frequency/Percentage*	*Frequency/Percentage*
Temporary	66.67%	95.83%	83.33%
Permanent	33.33%	4.17%	16.66%
Total	18	24	42

Table 11.6 Worker's type of contract by gender

Type of Contract	*Male*	*Female*	*Total*
	Frequency/Percentage	*Frequency/Percentage*	*Frequency/Percentage*
Oral	33.33%	50%	38.88%
Written	66.66%	50%	61.11%
Total	18	24	36

Permanently seasonal workers: informality versus instability

Workers' histories of employment, for the year in which they were interviewed, permit us to observe the existence of continuous employment in asparagus fields throughout most of the year. Following their histories of employment, more than half of workers interviewed have had a constant labor relation with asparagus farms in the year when the field research was conducted. For instance, when asking asparagus laborers about the time they spend working in asparagus fields in the year of the interview, at least 75% of the workers reported[6] that they were working at least seven or more months in asparagus fields, within which 61% reported working more than nine months. As Margarita pointed out, "[In April] I have also harvested, but I have harvested most of the time, the whole time I have been harvesting. This year, I had spent [it harvesting]" (W05, female, small farm, married, 3 children).

Moreover, as shown in Table 11.7, when examining the general time that participants worked in asparagus fields, an important group of workers reported

Table 11.7 Time working in asparagus farms in general

Type of Contract	*Percent*
Less than one (1) year	7.9%
Between one (1) and five (5) years	21.1%
Between six (6) and then (10) years	31.6%
Between 11 and 15 years	21.1%
More than 15 years	18.4%
Total	38

being employed repetitively through the years (with intermissions) in asparagus farming for as long as they can remember. For instance, 64.3% of the workers interviewed reported to have worked in asparagus fields, independently of scale of production, for six or more years and 39.5% more than 11 years. With the longest labor commitment, Juan reported to be working more than 30 years on asparagus farms. As Juan stated, "(. . .) at asparagus, let's see. If I am forty-seven, at asparagus I am at least thirty-five years. Thirty years in asparagus, at least thirty years working" (W20, male, medium farm, separated, 2 children). Although the interviews did not account for possible intermissions or breaks in workers' histories of employment, the long and constant commitment to their work in asparagus fields illustrates workers' status as a continuous labor force of this industry despite the existence of informal and temporal contracts with one farm.

The existence of a continuous labor relation through the years does not necessarily mean that laborers have or have had permanent contract relations with one asparagus producer or one asparagus farm. On the contrary, when examining the rotation patterns among laborers interviewed, workers had not worked exclusively at one farm over time or in a scale of production. Regarding their employment rotation patterns in 2015, 70% reported working for more than one farm in that year, within which 30% worked on at least three or more farms. Workers worked for both small and large farms and moved among different growers during the year to secure their employment in the asparagus fields.

The existence of rotation patterns, which are also linked to the employment status, scale of production and gender of the laborers, appears more clearly when examining the organization of labor and the existence of labor positions in asparagus agricultural fields.

Labor positions and gender division of labor

As can be seen in Table 11.8, the most common job position among workers was the "cortador o cosechador de espárragos [asparagus cutter or harvester]," with half of workers reporting working under this job position. Workers employed as asparagus cutters or harvesters encounter high levels of job availability during harvest time, since there is a great demand for labor power in this season.

Table 11.8 Worker's job position by gender

Job Position (most recent employment)	*Male*	*Female*	*Total*
	Percentage	*Percentage*	*Percentage*
Asparagus cutter/harvester	22.22%	70.833%	50%
Day laborer/ General laborer	5.55%	25%	16.66%
Field collector and packager	22.22%	0	9.52%
Field evaluator	5.55%	0	2.38%
Field security guard	5.55%	0	2.38%
Irrigation specialist/ water-well engine operator	16.66%	0	7.14%
Pest control/Pesticides applicator	11.11%	4.17%	7.14%
Right-hand man	11.11%	0	4.76%
Total	18	24	42

Despite the existence of many opportunities, to access a job as a cutter, workers do not work constantly on a single farm. If asparagus cutters or harvesters decide to continue working in the same position, they must usually find a job with another farm and continue rotating from farm to farm. This was observed especially for workers of small farms; the system of rotation is continuous from one producer to the other. As Raúl explained:

> Look, the harvest normally, didn't I tell you? They[the harvest] used to last three months (. . .) now [the harvest] minimum lasts two months. From there to two months, and sometimes [the producer] close a field, as he is working an average of five hectares. [He] kill [harvest] three [hectares] and two [hectares] later. For what? So, that we do not lack the work.
>
> (W03, male, small farm, married, 3 children)

In this sense, there appears to be a silent consensus among small producers to produce in different months, thereby allowing workers to move around from farm to farm. This situation is convenient not only for workers but also for producers who will not have to fight to ensure the labor power needed for their production. As Graciela explained, "(. . .) I've been working there, but it was eventually, they laid me off – that is, for 6 months the harvest was over, they laid me off. I went to work elsewhere and then when the season started, I went back there" (W07, female, large farm, married, 2 children).

Large corporations have a longer period of harvest time since their larger scale of land enables large-scale production, with many plots operating simultaneously. Moreover, the large scope of farms at this scale of production requires laborers to perform other activities besides harvesting such as planting, watering or cleaning. When there are no job opportunities as asparagus cutters, laborers can perform other tasks in asparagus fields.

As shown in Table 11.8, among the laborers interviewed, job positions other than asparagus cutters and asparagus day laborers include day field collectors and packagers, pesticides applicators or pest controllers, field security guards, irrigation specialists or water-well engine operators, right-hand men and field evaluators. When examining gender differences in job positions, there is a clear gender division of labor among workers interviewed. For instance, female workers mostly performed jobs as asparagus cutters or day laborers. On the contrary, male workers reported performing a wide range of jobs such as day field collectors and packagers, pesticides applicators or pest controllers, field security guards, irrigation specialists or water-well engine operators, right-hand men and field evaluators. These job positions usually respond to jobs that require specific skills, are required throughout the year or require more trust. Thus, this gives men more opportunities to hold a job in the off-season as well as to hold job positions at a higher pay.

Working shifts: there is no rest at harvest time

Asparagus laborers reported a wide range in the number of working shifts (see Figure 11.1). The variations in the working shifts among workers were related to their time commitment to their jobs, their job positions and the season of asparagus production. The research identified at least two different working shifts among full-time workers interviewed (74% of the workers interviewed), the continuous shift and the split shift.

The continuous shift is from 6:00 a.m. to 1 or 2 p.m. and generally occurs during the off-season, when the asparagus plant is growing, and the asparagus does not need to be harvested every day. During this season, workers, as day laborers, perform different type of activities such as weeding, planting, watering the plants, cleaning the fields, applying pesticides or controlling pests. According to workers, working under continuous shift is intense but allows them to leave the asparagus farm early and have the rest of the day for themselves. Rosa described how a day under continuous shift looks:

> Once at work, we arrive, we have a control, we have breakfast because there is a dining room, a dining room where we sit and everything. We have breakfast until the field manager, who is there the manager tells us "you know ladies that today we are going to carry out this work" (. . .) "Already girls, I have sent them to call you because I need [you] to pulled ribbons or I need [you] to weed." Sometimes there is [a call] for support also during harvesting [time] because when there is el reventón [the blowout] they need support also. (. . .) Sometimes the engineer says "now, [I will] give [you] tasks to all," so many "lines" and so many "lines." We do we kill [ourselves], we do not go to the bathroom to finish the task quickly, to get to the house.
>
> (W1, female, large farm, married, 4 children)

In a different manner, the split shift generally occurs during harvest time from 6:30 a.m. to 11:30 a.m., and then from 2:00 to 5:00 p.m. approximately. According to workers, the asparagus is a crop like no other since it requires being cut (or picked up) every day and all day long during the sunlight hours. The asparagus plant grows by sunlight every hour, and therefore it needs to be cut at a very precise moment, when it reaches certain length (largely due to commercialization specifications). If, for any reason, the asparagus plant grows beyond the length required, it will not be useful to be commercialized. As Jorge described, "[Asparagus] harvest cannot remain from one day to another, [it] cannot! [We] have to finish [it]. Whatever [the cost], [we have to] harvest!" (W31, male, large farm, married, 2 children).

What does this mean for the organization of labor in the fields? First, it requires that the asparagus is cut every day (the seven days of the week) through the daytime and second that laborers finish cutting all asparagus plants in the fields that have reached the correct size. Why do workers stop at the middle of the day? As Julio explained, "The people are resting in the dressing room during this time because here we stay, they do not go to the field, because in the field it burns in this time [of the year]" (W38.1, male, large farm, married, 2 children). Asparagus needs high levels of sunlight to grow. In this regard, laborers report that it is almost impossible to work directly under the sunlight during the middle of day due to high levels of sunlight in the region,[7] especially during the summertime. After the break, workers return to the work shift until 5 p.m. or 6 p.m., depending on the amount of asparagus in the fields for that day.

Workers emphasized the idea that the type of work that is demanded of them in the harvest season requires a full-time commitment, especially at large farms. This situation can limit their commitment to any other activity, and it will be necessary to schedule a replacement if for some reason a worker will not be able to attend to their shift. As we can observe in the case of Irene:

> The bad thing when you work in the field is that you cannot miss a day. (. . .) They do not give you permission, they do not even pay you the day. [In other words,] if you do not go one day, you have to look, to see the way who you will send to cut for you. Even in your birthday, you are ill, you have to go or send your replacement. Even if you miss a day, [they] do not pay the day, [it will be a] day lost.
>
> (W08, female, medium farm, separated, 5 children)

As Irene explained above, there is not a day off or holiday during the harvest period, and workers acknowledged this by expressing that "since the asparagus do not rest, they cannot rest either." According to workers, as Jorge pointed out before, missing one day of work generates problems for asparagus producers, particularly in losing out on potential asparagus production if the crop is not harvested at the precise time. Therefore, workers cannot miss a day of work. During the period of "el reventón [the blowout]," what workers define as the period of major asparagus production soon after the harvest season starts, workers usually work past their shift and do not know exactly at what time their shift will end. Additionally, during the harvest season, workers are required to work seven days per week from the early morning to the end of the day. For instance, more than 65% of the workers reported working (or having worked previously in their most recent job in asparagus fields) seven days per week, while 30% work six days per week. This situation goes in line with Juana's experience, who reported to not just work more intensively but with overtime as well during the harvest season:

> When there is harvest, it is every day. From Monday to Sunday, the same schedule. Sometimes even you leave [the field] at 6 p.m., at 6:15 p.m. because of the harvest. You have to leave depending on the time you finish harvesting. (. . .) Sometimes you leave [for break] at 12:30 p.m. You do not have an exact schedule, you leave [but] if there is [harvest] remain, no! You have to finish the harvest and at the hour that you finish, you leave. In the afternoon, if you finish the harvest, you go home.
>
> (W35, female, large farm, married, 2 children)

A third working shift is the overnight shift, although it is less common than the continuous and split shifts. About 9.5% of laborers interviewed noted working all night long, from 5:00 p.m. to 5:00 a.m. Workers in the overnight shift performed different job positions such as being a security guard, an irrigation specialist or a water-well machine operator. In such cases, the laborers worked over eight hours, generally a 12 hour-shift. In some cases, the overnight shift is complemented with a part-time shift during the harvest time, as is the case of Raúl, who works as right-hand man in a small asparagus farm, working in the morning time as field packager, then returning to the farm in the evening to stay all night as a security guard. Of the group of workers who reported working an overnight shift, all were men workers with permanent employment status.

A last working shift identified among sampled workers is the part-time shift, a three- to five-hour work shift. The part-time shift is more prevalent in small asparagus farms. According to the interviewees, there are two main types of part-time work shifts, the morning shift and the afternoon shift. The morning shift "is from 6:00 a.m. to 10:30 a.m., which is the normal schedule" but can be extended until 11–11:30 a.m. (W06, Estrella, female, small farm, long-distance relationship, 1 children), while in the afternoon shift, "the cut [harvest] starts from 2 in the afternoon (. . .) until 5" (W12, Natalia, female, small farm, separated, 1 children). Although the morning is the most common time to harvest asparagus among smallholders, there are also smallholders that harvest in the afternoon. When examining gender differences, women make up the majority of part-time workers in this study sample. There is a group of workers that work seven to eight hours per day by combining two part-time working shifts in small asparagus farms, the hours are similar as the split-shift in the large farms but they are employed by two different producers.

To sum up, workers' schedules appear to be in constant change throughout the different seasons of production and depending on their job positions. In this instance, workers not only change or rotate farms as described before but change working shifts between seasons, with most workers reporting to work with more intensity during the harvest season than the off-season.

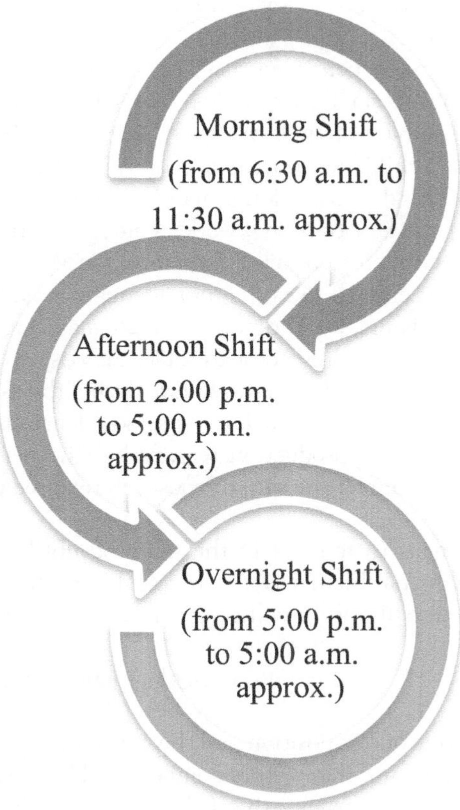

Figure 11.1 Working shifts in the harvest season

Salaries: it pays to eat but it is not enough for more

Workers' salaries differ by job position, employment status and the season of production. Although it is not possible to define one amount, it is possible to present some results from workers' responses. In the first place, for all the workers interviewed, the salary reported by workers was over the monthly minimum wage established by the national government[8] when the field research was conducted but below the living wage established by national labor unions. Workers report that this salary is higher than other non–agricultural-related industries, especially in the harvest season. However, there is a salary gap by scale of production between small, medium and large farms that corresponds with the type of working shift and the number of hours.

In the harvest season, asparagus cutters appear to earn substantially more than workers working as day laborers, who fulfill this position generally during the off season. In the same regard, workers report earning higher salaries on large asparagus farms than workers on small farms. The net salary of small farms is considerably less than the net salary of the large farms; however, it corresponds to a shorter work shift.

Although some workers perceived that large farms during harvest time "pays well" in comparison with small farms (W04, Marisol, small farm, married, 3 children) and during the harvest time the salary "does provide" (W26, Mercedes, large farm, married, 3 children), at least 50% of the workers noted dissatisfaction with their salaries. This dissatisfaction is mainly related to two reasons.

First, in terms of the cost of living, workers argue that this salary –especially in the off-season – is not enough to fulfill their responsibilities. As, Jimmy stated that their salary is just enough to eat but not adequate to go beyond that, it only provides "for the weekly life, we work to eat" (W25, male, medium farm, married, 1 child). A second contributing factor to salary dissatisfaction is related to the intensity of labor that the workers must endure during the asparagus harvest. As Margarita states, the work during the harvest season demands not only asparagus cutters' commitment to all production seasons but the performance of longer hours "crouched down" under the sun, the position required to cut asparagus:

> Well the asparagus is from six in the morning until eleven o'clock. And the heat, it gives you thirst. You are crouched from the time you enter until eleven o'clock and also it hurts the waist because, it hurts. It is a killing job as they said.
>
> (W05, female, small farm, married, 3 children)

The negative effects of this position and the constant exposure to the sunlight are considered work hazards that are not compensated by the salary workers earn. In their narratives, workers remark that fieldwork cannot be compared to other types of work, such as in an office or factory, in which workers are protected by a roof and are not required to be "crouched down."

Labor benefits

The scale of production, the type of contract and the employment status are perhaps the deciding factors that determine which workers receive or do not receive labor benefits. As can be seen in Table 11.9, around 54% reported not receiving social benefits in their current (or most current) jobs, while 42% reported receiving them.

Workers reported that social benefits are out of the question when working for asparagus smallholders, who argue that the labor benefits are included in workers' daily salary, a salary that is comparatively higher than the salary of large farms. Workers with formal or written contracts, who generally work for large farms, have access to labor benefits. The length of employment and employment status (temporary or permanent) will define when and what type of benefits workers have access to. When observing gender differences, female workers reported having less access to social benefits, which goes in line with their temporary job positions as asparagus cutters and their higher employment rates on small farms.

Table 11.9 Worker's social benefits by gender

Social Benefits	*Male*	*Female*	*Total*
No	38.88%	66.66%	54.76%
Yes	61.11%	29.16%	42.86%
Missing	0	4.16%	2.38%
Total	18	24	42

Job satisfaction

Despite the depiction of the work as intensive or sometimes a "killing job" under the sun, workers also value different aspects of their labor at asparagus farms. Perhaps the most valuable aspect reported by workers related to the existence of friendships bonds that made this experience a "cool" one. The perception that working in agricultural fields has a cool or fun side also relates with the value that some workers like Nikole give to the work on agricultural fields. For example, "you go there with your friends, talking, everything calm, you tell your things. It's fun, cool, the hour fly. Instead of being in a position [such as] selling clothes [it is] boring" (W09, Nikole, female, large farm, separated, 1 child).

Furthermore, workers' narratives of labor satisfaction need to be examined in relation to their employment on small, medium or large farms. With dissimilar perceptions of their job benefits, workers employed at small farms valued the existence of more horizontal relations with other co-workers and the producer. They perceived that they carry out their work more independently, having more flexibility not only to do their work but to manage their time and personal commitments. This independence, however, has a high cost according to workers employed in large asparagus farm: the lack of benefits. Raúl and Rosa expressed their preferences to work in small and large asparagus farms.

> They treat you badly, they treated you badly. In [a large farm], down there like they treat you badly or sometimes they yell at you and I'm not one of those. I do not like to be boos. I'm better independent and I work my hours, what it is and [then] I rest. Calm, nobody is bugging me now.
>
> (W03, male, small farm, married, 3 children)

> In the large farm, you have your benefits (. . .) On the other hand, in parcels, you only [get] pay your daily wage. There you are not paid insurance, you are not paid, you are not deducted from the AFP [Social Security], you are not paid your allowance, you are not paid any benefit, which is what the large farm does.
>
> (W01, female, large farm, separated, 4 children)

Conclusions and discussion

The study in two towns dedicated to the production of asparagus confirms the existence of an industrial production of asparagus for export, production that demands a large contingent of agricultural workers. The research in these major asparagus-producing towns brings to light processes of modernization and industrialization of agriculture in the Global South, as well as describes the situation of asparagus agricultural wage workers as part of the proletarianization of the peasantry examined by different scholars.

Agricultural wage workers in this study are understood in relation to their contractual relationships, as mediated by salary, with the owners of the asparagus farms in which they are employed. For these workers, their main source of income was the salary they received for the agricultural work they performed on someone else's asparagus farms. It is important to note that the entire sample worked in asparagus fields at some point during 2015, with agricultural labor being their primary job.

Research findings suggest that the notions of temporary and flexible labor appear to be limited in understanding asparagus workers' history of employment and commitment to the asparagus industry in this study. This research argues that despite the characterization by the literature and national legislation (Law 27360) of the industry's categorization of employment as seasonal and asparagus wage workers as temporary labor force, most workers in this study sample find themselves working continuously in asparagus farms.

For instance, workers report that because of the continuous rotation among asparagus farms, they have consistently worked throughout much of 2015, as they have in previous years. It is important to note that the use of categories of temporary and permanent workers do reflect in part the seasonal nature of asparagus production and agricultural production in general. However, a strict distinction of both categories – temporary and permanent – might be problematic when understanding the type of labor participation of this contingent of workers. In this sense, although asparagus workers labor contracts, informal or formal, defined them as a temporary workforce, this research findings suggest that this characterization might obscure workers' continuous and intense commitment to the industry. Furthermore, research findings suggest that workers' status as temporary in juxtaposition with their ongoing commitment to the industry might lead to labor informality, instability and flexibility. The development of informal contracts in small-scale production and the implementation of Law 27360 in large-scale production are examples of the limitations placed on workers' benefits that can result from the mischaracterization of workers' type of participation.

Regarding working conditions, this study's findings suggest that workers employed in asparagus farms are still subject to conditions of vulnerability, which previous studies about the industry also suggest. Among the critical concerns related to working conditions are the limited access and unavailability of benefits because of their categorization as temporary workers. A second element connected to the vulnerability of employment is related to the producers' demand for high levels of commitment from workers, in terms of time (number of hours and days) and work intensity. Once again, despite the categorization of this employment as temporary, workers are demanded at harvest time to maintain steady jobs. This means working seven days per week under stable schedules and performing physical work under extreme climate conditions that leads to permanent physical wear.

Additionally, the research finds the existence of low levels of satisfaction with the salaries earned in the asparagus industry, despite the fact that this salary hovers around – and in harvest time, above – the national minimum wage. Workers in this study pointed out that their earnings are not commensurate with the intensity of the work performed on asparagus farms, as well as how these earnings do not allow them to go beyond their basic necessities. This corresponds with the empirical work about Peruvian agricultural workers and the discussion presented by Moura (2010) about the existence of a gap between the minimum wage and living wage, which for the author leads to agricultural workers being defined as the working poor. The existence of low salaries in the asparagus industry, and the country in general, can also be related to literature regarding the existence of the international division of labor, in which Global South nations act as suppliers of cheap labor (de Janvry and LeVeen, 1986). This access to cheap labor is one of the comparative advantages to developed industries, such as the existence of NTAEs (Paske, 1996; Villachica and Toledo, 2003, p. 65; Meade et al., 2010).

Based on the narratives of workers, labor satisfaction relates to the type and scale of production. On the one hand, workers employed at small farms relate job satisfaction to the positive environment of their workplace, which they describe as featuring close labor relations between laborers and asparagus producers. Also noted is the considerable autonomy while doing their agricultural labor. On the other hand, asparagus laborers in large farms relate job satisfaction to the large amount of jobs provided by the asparagus farms in the region, as well as the benefits offered under formal contracts.

Empirical studies conducted in Peru regarding the labor organization of the asparagus industry point to the existence of a gender division of labor in both asparagus fields and processing plants. In the same line to other studies, this study found a gender division of labor among asparagus workers. For instance, women work almost exclusively as asparagus cutters in contrast to men, who appear to have more employment options. The seasonal nature of the asparagus cutter's job increases the flexibility and high rotation of employment for women. Thus, often, women hold fewer stable jobs and therefore have less access to labor benefits and rights. This situation may lead to the vulnerability of working conditions, as other authors studying industrial agriculture in Peru and Latin America have suggested (Soto Baquero and Klein, 2012a; Soto Baquero and Klein, 2012b).

When observing workers' perceptions about their own work (and not necessarily from the industry's interests), the notion of "disposability" (Wright, 1999) appears to take on new nuances. In this sense, for example, workers appear to perceive their labor and participation in the industry as irreplaceable instead of "disposable." The notion of irreplaceability is derived from the understanding that the asparagus industry is dependent upon the existence of a stable and reliable contingent of workers. In this sense, the asparagus industry not only wears out workers but depends on them to function. High numbers of them reported working for many years in asparagus farms. More importantly, workers acknowledge the central role of their labor by acknowledging the negative impacts for asparagus production if they miss a day of work. They also point out the lack of consistency between their salaries and the value of their labor. In this regard, this study argues that the use of irreplaceable workers can also highlight workers' own perceptions of the value of their labor participation and, with that, overcome the victim approach as suggested by third-world and transnational feminists (Mohanty, 2003).

By acknowledging and giving value to the high commitment that workers have to their work in asparagus farms, in quantitative and qualitative terms, this study might help to develop a new category of employment labor relations that might understand this relation as a continuum. In addition, this study can provide valuable information to policy makers that will allow them to first gain a better understanding of the differences and nuances of the labor relationships and statuses of workers employed in this industry, as in the case of the workers who are also permanent residents in the asparagus-producing areas.

Notes

1 For example, the Millennium Goal N.1. Target 1. B seeks to achieve "full and productive employment and decent work for all, including women and young people" (United Nations, 2015, p. 17).
2 Piura, Lambayeque, La Libertad, Ancash, Lima and Ica.
3 Farms size exceeds 5 hectares and is smaller than 80 hectares.
4 Among them are at least two important agro-industrial firms at the regional and national levels.
5 In the case of workers who work in both types of scales of production, when analyzing workers' labor experiences in asparagus farms, this study only examined the labor experience of the primary source of employment.
6 This information was registered for 37 workers, with 32 workers reporting working seven or more months in asparagus farms.
7 Which have also the high level of solar radiation.
8 When the interviews were conducted, the national minimum wage was 750 nuevos soles [231 U.S. dollars at current exchange rate], however the government raised the minimum wage up to 850 soles [261 U.S. dollars] on May 1, 2016 (Ollanta Humala, 2016).

Works cited

Allen, P. and Sachs, C. (2007). Women and food chains: The gendered politics of food. *International Journal of Sociology of Food and Agriculture*, 15(1), pp. 1–23.

Appendini, K. (2002). From where have all the flowers come? Women workers in Mexico's nontraditional markets. In: S. Razavi, ed., *Shifting burdens: Gender and agrarian change under neoliberalism*. Bloomfield, CT: Kumarian Press, pp. 93–108.

Ballara, M. and Parada, S. (2009). *El empleo de las mujeres rurales. Lo que dicen las cifras. [Employment of rural women. What the figures say]*. Santiago de Chile: FAO – CEPAL.

Benson, B. (2008). 2005 Update of the world's asparagus production areas, spear utilization, yields and production period. *Acta Horticulturae*, 776, pp. 495–508.

Burawoy, M. (1991). The extended case method. In *Ethnography unbound: Power and resistance in the modern metropolis*. Berkeley and Los Angeles: University of California Press, pp. 271–300.

Chacaltana, J. and Yamada, G. (2009). *Calidad del empleo y productividad laboral en el Perú [Quality of employment and labor productivity in Peru]*. Working paper, 691. Washington, DC: Inter-American Development Bank. Available at: https://publications.iadb.org/bitstream/handle/11319/2177/Calidad%20del%20empleo%20y%20productividad%20laboral%20en%20el%20Per%C3%BA.pdf?sequence=1&isAllowed=y.

Collins, J. (1995). Gender and cheap labor in Agriculture. In: P. McMichael, ed., *Food and agrarian orders in the world-economy*. Westport, CT: Greenwood Press.

Deere, C. (2005). *The feminization of agriculture? Economic restructuring in rural Latin America*. Occasional Paper, 1. United Nations Research Institute for Social Development. Available at: www.unrisd.org/80256B3C005BCCF9/(httpP ublications)/20024EBC6AB9DA45C1256FE10045B101?OpenDocument.

de Janvry, A. and LeVeen, E. P. (1986). Historical forces that have shaped world agriculture: A structural perspective. In K. A. Dahlberg (Ed.). *New directions for agriculture and agricultural research: Neglected dimensions and emerging alternatives*. Totowa, NJ: Rowman and Allanheld, pp. 83–104.

Díaz Rios, L. (2007). *Agro-industries characterization and appraisal: Asparagus in Peru*. Agricultural, Management, Marketing and Finance Working Document, 23. Rome: FAO. Available at: www.fao.org/docrep/016/ap297e/ap297e.pdf.

Federación de Mujeres de Ica [FEPROMU – ICA]. (2007). *Estudio de Aproximación. La otra cara de la agroindustria en Ica: los efectos en la salud de las trabajadoras [At first glance study. The other side of agroindustry in Ica: The effects on the health of women workers]*. Lima: FEPROMU-ICA.

Ferm, N. (2008). Non-traditional agricultural export industries: Conditions for women workers in Colombia and Peru. *Gender and Development*, 16(1), pp. 13–26.

Flores Mego, K. (2011). La agroexportación no tradicional en el país de las maravillas. Condiciones de trabajo y derechos laborales de las mujeres [Non-traditional agroexports in wonderland. Working conditions and labor rights of women]. In: CEPES, eds., *Mujer Rural: Cambios y Persistencias en América Latina [Rural women: Changes and persistences in Latin America]*. Lima: CEPES, pp. 117–144.

Food and Agricultural Organization of the United Nations. (2017). *Food and agriculture data[FAOSTATS]*. Rome, Italy: FAO. Available at: www.fao.org/faostat/en/.

Little, P. (1994). Contract farming and the development question. In: P.D. Little and M. Watts, eds., *Living under contract: Contract farming and agrarian transformation in Sub-Saharan Africa*. Madison: The University of Wisconsin Press, pp. 216–247.

Marañón, B. (1993). *Obreros en la industria esparraguera: valles de Chao Virú e Ica. [Workers in the asparagus industry: valleys of Chao Virú and Ica]. Revista Debate Agrario. Análisis y alternativas*. Lima: Debate Agrario.

Meade, B., Baldwin, K. and Calvin, L. (2010). *Peru: An emerging exporter of fruits and vegetables (A report from the economic research service)*. Washington, DC: United States Department of Agriculture.

Mohanty, C. (2003). *Feminism without borders: Decolonizing theory, practicing solidarity*. Durham and London: Duke University Press.

Moura, F. (2010). *Prácticas discriminatorias en las relaciones socio laborales en la agro exportación del espárrago en Ica, Perú: 1999–2009 [Discriminatory practices in the socio-labor relations in the agro export of asparagus in Ica, Peru: 1999–2009]*. Tesis para obtener el Título de Magister en Sociología. Unpublished Master thesis. Lima: PUCP.

Nichols, M. (1990). Asparagus the world scene. *Acta Horticulturae*, 271, pp. 25–31.

Ollanta Humala decretó aumento del sueldo mínimo a S/850 [Ollanta Humala decreed increase of the minimum salary to S / 850]. (2016). *El Comercio*, March 30. Available at: http://elcomercio.pe/buscar/?query=sueldo+minimo+ollanta+750.

Paske, M.R.A. (1996). Importing fresh asparagus. A personal view. *Acta Horticulturae*, (415), pp. 19–24.

Patel-Campillo, A. (2012). The gendered production – consumption relation: Accounting for employment and socioeconomic hierarchies in the Colombian cut flower global commodity chain. *Sociologia Ruralis*, 52(3), pp. 272–293.

Patel-Campillo, A. (2011). Transforming global commodity chains: Actor strategies, regulation, and competitive relations in the Dutch cut flower sector. *Economic Geography*, 87(1), pp. 79–99.

Preibisch, K. and Grez, E. (2010). The other side of el Otro Lado: Mexican migrant women and labor flexibility in Canadian agriculture. *Signs: Journal of Women in Culture and Society*, 35(2), pp. 289–316.

Radel, C., Schmook, B., Mcevoy, J., Mendez, C. and Petrzelka, P. (2012). Labour migration and gendered agricultural relations: The feminization of agriculture in the Ejidal sector of Calakmul, Mexico. *Journal of Agrarian Change*, 12(1), pp. 98–119.

Ruiz-Bravo López, P. and Castro Bernardini, M. (2012). Caso de Perú [Case of Peru]. In: F. Soto Baquero and E. Kelin, Coord., *Empleo y condiciones de trabajo de mujeres temporeras agrícolas. Tome 2. [Employment and working conditions of agricultural temporary women. Tomo 2]*. Rome, Italy: FAO, pp. 194–268.

Sachs, C. and Alston, M. (2010). Global shifts, sedimentations, and imaginaries: An introduction to the special issue on women and agriculture. *Signs: Journal of Women in Culture and Society*, 35(2), pp. 277–287.

Singleton, Jr. R. and Straits, B. (2010). *Approaches to social research*. 5th ed. New York and Oxford: Oxford University Press.

Soto Baquero, F. and Klein, E. (Coord.) (2012a). *Empleo y condiciones de trabajo de mujeres temporeras agrícolas [Employment and working conditions of temporary agricultural women]*. Tome 1. Rome, Italy: FAO-CEPAL – OIT.

Soto Baquero, F. and Klein, E. (Coord.) (2012b). *Empleo y condiciones de trabajo de mujeres temporeras agrícolas. [Employment and working conditions of temporary agricultural women]*. Tome 2. Rome, Italy: FAO-CEPAL – OIT.

United Nations. (2015). *The millennium development goals report*. New York: UN.

Válcarcel Carnero, M. (2003). *Nuevas relaciones sociales entre los productores, la industria agroexportadora y las ONG en el sector agrario peruano. Estudio del sistema de producción de los espárragos entre 1980 y 2000 [New social relations between producers, the agro-export industry and NGOs in the Peruvian agrarian sector. Study of the production system of asparagus between 1980 and 2000.]*. Dissertation. Belgium: Catholic University of Leuven.

Wright, M.W. (1999). The dialectics of still life: Murder, women and maquilladoras. *Public Culture*, 11, pp. 453–473.

12 Gender equality and trees on farms

Considerations for implementation of climate-smart agriculture

Tatiana Gumucio, Diksha Arora, Jennifer Twyman, Ann R. Tickamyer and Mónica Clavijo

Research problem

The use of trees on farms has been promoted as an important climate-smart agricultural practice in various parts of the world most vulnerable to the impacts of climate change. Climate-smart agriculture (CSA) is an approach with three primary objectives: increased smallholder productivity, enhanced resilience to climate change and reduced greenhouse gas emissions. While CSA often emerges as a priority strategy for climate change initiatives seeking to promote smallholder resilience, it is critical that plans consider how gender and social inequalities can inhibit significant portions of the population from participating in and benefiting from resilience-building efforts. While both women and men are integral players in natural resource management, norms and institutions surrounding gender roles and responsibilities can promote inequalities in access to land, capital and technical services, to women's disadvantage (FAO, 2011). Correspondingly, men and women confront varying opportunities and challenges to engage with and implement agricultural innovation, particularly in a context of climate variability and change (Jost et al., 2016; Kristjanson et al., 2017). Policies and programs that do not take into account gender-based challenges to access and benefit risk inadvertently creating harm to women (Kabeer and Subrahmanian, 1996). Consequently, consideration of gender equality in CSA initiatives is important to ensure and advance women's human rights and empowerment (Asher and Sijapati Basnett, 2016; Bee and Sijapati Basnett, 2016; Doss, 2017); it is critical in order to achieve impacts at scale as well (Kantor et al., 2015).

The chapter analyzes the factors that can enhance household capacities to implement CSA practices related to trees on farms in a territory identified as vulnerable to climate change impacts in Nicaragua, through an empowerment lens. Specifically, we pose the following research questions: what key factors contribute to increased household implementation of CSA practices related to trees on farms? More particularly, how are aspects of women's empowerment part of the enabling environment? The study's approach to CSA implementation is particularly important in that it does not focus on the behavior and characteristics of the household head, as other adoption studies tend to do, but rather household primary decision makers. In this way, it is possible to assess the influence of female primary decision makers living in male-headed households on household adoption. We address the questions and elaborate the study in five sections. We first present a brief review of the literature on how gender equality and aspects of women's empowerment can contribute to implementation of agroforestry and sustainable agricultural practices and subsequently explain the theoretical framework guiding the study. After a description of the mixed-methods research design, we discuss the results from a probit regression model and semistructured interviews. We conclude with recommendations for the development of policies and interventions targeting smallholder agriculture that do not reinforce but rather transform limiting gender inequalities.

Literature review

While further research is necessary, existing studies on gender considerations in climate change adaptation and food security interventions suggest that promoting women's empowerment can enhance the impacts on household livelihoods and production. For example, research in South Vietnam demonstrates that increasing women's access to trainings on climate-smart technologies, such as stress-tolerant seeds and associated crop management, can allow women to make more informed decisions on rice production, increasing yields in some situations (Chi et al., 2015). Moreover, a study in Malawi suggests that joint participation in household decision making between spouses can enhance financial and livelihood planning (Farnworth et al., 2018). Critically, the research highlights that women's agency can be an important factor for household well-being, even in situations wherein household access to resources is limited.

Other research demonstrates how women's empowerment may contribute to increased household agricultural innovation or implementation of CSA. For instance, consideration of gender differentials in access to resources, particularly agroforestry information, can be important for the implementation of CSA initiatives related to agroforestry. Catacutan and Naz (2015) highlight gender inequalities that disadvantage women with respect to land, labor, capital and agroforestry knowledge and skills; correspondingly, one of their key recommendations is that agroforestry interventions include women-specific extension services. Similarly, research on oil palm production in Papua New Guinea demonstrates that women can often face limited access to information and capacity building due in part to the targeting of men in extension programs that promote agroforestry systems (Koczberski, 2007).

Besides access to information, land control and decision-making power can be important to facilitate action on information received. For instance, several studies highlight that women's limited control of land and limited involvement in agricultural decision making can make climate information irrelevant for rural women (Roncoli et al., 2009; Carr, 2014; Poulsen et al., 2015; Carr et al., 2016; Carr and Owusu-Daaku, 2016; Serra and McKune, 2016; Carr and Onzere, 2017). However, while fewer women than men tend to have land rights, research in West Bengal, India, demonstrates that when women's names are included in formal land titles, households are more likely to include women in agricultural and food decision making (Santos et al., 2013).

Additionally, research on women's participation in producer organizations and community groups demonstrates the benefits women can perceive in terms of improved leadership skills, social networks and access to capacity building. In general, group participation can facilitate women's access to information important for agricultural and livelihood planning and create networks for dissemination of innovation (Pandolfelli et al., 2007, p. 5; Venkatasubramanian et al., 2014; Coulier, 2016). Concerning high-value tree crops like coffee, women members of producer organizations can experience increased acquisition of new knowledge through organization trainings and technical assistance (Jazibi Cárcamo et al., 2010; Higuchi et al., 2012; Twin, 2013). Studies of women and non-timber forest product (NTFP) producer organizations and community development projects in Latin America demonstrate that women access valuable trainings on sustainable NTFP development through their membership (da Cunha, 2006; Shreckenberg and Marshall, 2006;Mello, 2014). The studies consistently note that women gain enhanced self-esteem, confidence to seek information and leadership experience that they apply in household and extra-communal affairs as well.

While participation in community and producer groups can facilitate women's greater access to agricultural information and leadership development, it is important to consider that not all women may be able to participate. Several of the factors that influence women's exclusion from groups are related to restrictive gender norms and spousal disapproval of organization participation. For example, women may often feel pressured to refrain from joining NTFP producer unions or forest management groups due to spousal disapproval (Elias and Arora-Jonsson, 2016; Evans et al., 2016). Such discontent can stem from the concern that women's participation in public-sphere activities would undermine male household authority (Evans et al., 2016).

Further research suggests that in households where spouses practice more gender-equal relations, adoption of sustainable cash crop intensification practices like agroforestry may be more likely. A study in Uganda of spouses who had chosen to participate in a gender program that promoted equitable spousal decision making versus those who did not found that the households of those in the program more frequently adopted sustainable intensification practices for coffee production, such as mulching, stumping and planting of shade trees (Lecoutre and Jassogne, 2016). Correspondingly, other research suggests that more trees may tend to be planted in households where women contribute to decision making on tree planting and management (Meijer et al., 2015).

In this way, the literature suggests that intra-household gender equality and women's empowerment can influence implementation of climate-smart agricultural practices related to agroforestry and the use of trees on farms. More particularly, it suggests that factors such as access to agricultural information, land control, agency in decision making, group membership, and gender norms can influence women's – and, correspondingly, the household's – capacity to implement agroforestry and sustainable agriculture. The following section explains a theoretical framework related to women's empowerment used to structure the relations among the various factors identified for the study.

Theoretical approach and key concepts used

Our understanding of women's empowerment as it relates to CSA implementation is based on a combination of approaches, targeting women's empowerment in agriculture, gender equality in agricultural innovation and gender-transformative change. The study draws from the Women's Empowerment in Agriculture Index (WEIA), a tool developed by the International Food Policy Research Institute (IFPRI) and used by USAID in Feed the Future countries, in order to conceptualize certain dimensions of women's empowerment. The framework highlights the capacity to make decisions and the resources (material, social, human) to act on them as key to empowerment (Kabeer, 1999). Correspondingly, the WEIA identifies five domains of empowerment; for the purposes of the study and due to data limitations, the chapter focuses on three of these: production, resources and leadership (Alkire et al., 2013).[1] The production domain addresses the actor's capacity to participate in decision making on agricultural production. The resources domain concerns control over resources necessary to act on those agricultural decisions; consequently, it analyzes access and control over critical productive resources, as well as ability to make decisions on those resources. Finally, the leadership domain measures aspects of inclusion, participation, and organizational capacity.

While the WEIA is useful, it can fail to address how the normative structures surrounding gender can condition women's capacities to act as well as the resources available to them. Analytical concepts for understanding how gender norms influence women's and men's capacities to implement CSA practices are drawn from the Enabling Gender Equality in Agricultural and Environmental Innovation[2] (GENNOVATE) project. The framework envisions an opportunity structure wherein institutions and gender norms varyingly condition actors' abilities to access and act on available resources and technologies for agriculture and natural resource management. In this way, norms can influence actors' capacities to exercise agency and innovate. At the same time, agency and innovation can cause shifts in the opportunity structure, creating change in gender norms as well (Badstue et al., 2014).

Our study also recognizes that transformative change, which addresses the root causes of gender inequalities, must act at both systemic and individual levels; furthermore, it involves change in both material and normative aspects (Cornwall, 2016). Sociocultural norms surrounding gender critically influence women's degree of empowerment; however, how actors accept, engage with, and resist such norms at an individual level are an important facet of change processes. In concert with normative change, shifts in material aspects involving access and control of resources are also a key consideration for promoting women's empowerment.

The study seeks to analyze how women's empowerment indicators – specifically, access to productive resources, group membership, agency in decision-making and gender norms – can interact to influence implementation of CSA practices. Most importantly, the study applies a gender lens, critically analyzing how aspects of women's empowerment may affect implementation of CSA practices related to trees on farms. The following section details the mixed-methods approach used for the study, combining quantitative and qualitative research methods.

Research design

Data collection

Intra-household survey

The study uses data collected in 2015 through an intra-household survey of the CGIAR Research Program on Climate Change, Agriculture and Food Security (CCAFS) in Tuma la Dalia, Nicaragua. The area constitutes a CCAFS Climate-Smart Village, a territory distinguished by high climatic risk and wherein CCAFS partners have established strong links with local communities. It also coincides with the Nicaragua-Honduras Sentinel Landscape of the CGIAR Research Program on Forests, Trees and Agroforestry (FTA). Two hundred and seventy-one households were chosen through simple random sampling. The surveys were carried out with adult men and women who identified themselves as household members who were primarily responsible for agricultural decision making. In this analysis, we focus on a sub-sample of 253 households with an adult woman who is one of the primary decision makers.

In addition to personal- and household-level socioeconomic characteristics, the intra-household survey collected data related to agricultural and agroforestry activities and sex-disaggregated data about adoption of CSA practices. Survey questions address domains used for measurement by the Women's Empowerment in Agriculture Index (WEIA), for example, access to productive resources, group participation and agency in decision making.

Semi-structured interviews

The statistical analyses were supplemented by qualitative data collected from semistructured interviews. Fourteen semistructured interviews were carried out with women and men primary household decision makers from two communities included in the household survey. In order to probe intra-household gender dynamics, women and men from the same household were interviewed; in other words, one couple from each of seven households, total. In this way, seven women and seven men were individually interviewed. It was found that group membership did not exist widely across women and men survey participants. Correspondingly, as this was a variable of interest for the study, it became an important requirement for selecting the communities for the semistructured interviews: in one community, agricultural or natural resource management groups existed minimally; in the other, they existed to a larger extent. Five households were interviewed in the former and two in the latter. Interviews included questions addressing access to information on agriculture and agroforestry and on agroforestry knowledge and implementation, household members' participation in agricultural and agroforestry decision making, participation in local organizations, and household and community norms surrounding women's leadership. Information from the semistructured interviews was analyzed using Atlas.ti for themes related to knowledge and implementation of CSA practices; access to information; land control; group participation; and household and community gender norms. Although the sample size is limited, the qualitative data and analysis help to assess normative structures underlying empowerment, with a focus on individuals' acceptance and engagement with beliefs surrounding gender roles.

Data analysis

It is assumed that households' decision to adopt CSA practices is a function of household constraints and women's empowerment. The economic constraints as well as different domains of women's empowerment may vary across households, which may affect households' adoption of CSA practices. A probit regression model is used to analyze how variables related to women's empowerment and land use might influence implementation of CSA practices related to trees. We measure implementation of CSA practices at the household level. The dataset informs separate responses of the principal man and the principal woman on whether the practices of agroforestry, live fences and reforestation were undertaken by the household in the previous year. Given two responses and the fact that in some cases, these responses are discordant, we calculate adoption rate by considering the households in which both spouses agreed to have adopted at least one of the three practices. In addition, we include the households in which one spouse said yes while the other did not respond due to lack of knowledge about the practice. We assume no adoption in the households where one spouse responded positive implementation and the other responded no adoption of the practice. In case of households with one decision maker, we consider only one response to adoption of CSA practices. The dependent variable used in the regression model is a dummy variable that takes the value of 1 for adoption of at least 1 practice and 0 otherwise.

Several indicators of women's empowerment are included as covariates. It is worth highlighting that the focus on women primary decision makers is not common in comparison to the majority of other adoption studies, which use the gender of one household head for their models. By using indicators of empowerment of women primary decision makers, it is possible to examine the behavior of female primary decision makers in male-headed households. Women's participation in productive and commercial decisions is included to measure women's agency in the decision-making process. The dummy variable takes the value 1 if the woman participated in farm decision making individually or jointly with her spouse or other household member/s. To measure empowerment in the resources domain, women's access to information on the use of trees on farms is represented by a dummy variable that takes the value 1 if the woman respondent received information from an external source or from internal sources like household members or neighbors. Women's prior knowledge of the practice is also included in the access to information variable.[3] Also, women's joint or individual ownership of at least one land parcel is included as a measurement of resource control. Women's participation in groups or organizations working on agricultural, natural resource management or women's issues is included as a dummy variable to measure empowerment in the leadership domain.

At the household level we control for various economic and land-use characteristics. The literature on agroforestry adoption suggests that adoption of agroforestry practices may also be affected by a number of other factors (Ashraf et al., 2015; Dhakal et al., 2015; Conteh et al., 2016; Versteeg et al., 2017). For example, lack of formal land rights may inhibit implementation of agroforestry practices. To measure the relationship between land tenure and adoption, a variable that measures whether the household rents any parcels is included in the model. Off-farm income has also been seen to positively influence a household's capacity to implement agroforestry-related practices; with this in mind, a variable measuring whether any of the primary decision makers have worked off-farm over the past year is used.[4] Total cultivated farm size is included in the model as a proxy for wealth, considering that wealthier households will be more capable of implementing agroforestry-related practices.

Land use can also significantly affect implementation of practices related to trees on farms. Research carried out by the CGIAR Research Program on Forests, Trees and Agroforestry (FTA) in Waslala and in Tuma la Dalia, Nicaragua, demonstrates that parcels used for shade coffee and cacao production have the greatest tree species diversity and tree density, in comparison to parcels used for grain or livestock production or home gardens (Amores Contreras, 2015). Furthermore, as shade complexity diminishes, so does number of individual trees and species. With these issues in mind, variables related to percentage of farmland used for (a) coffee production, (b) basic grains and (c) home gardens are included in the model. Table 12.1 presents the various explanatory variables used in the probit regression, along with their expected relationships with the dependent variable. Individual-level controls like the respondents' age and education level (number of years in school) are included as well.

Results

Household characteristics of the study group

Table 12.2 presents a summary of characteristics of households included in the survey. On average, households have 1.78 parcels of land, and average parcel size is 2 hectares, while average farm size is 3.37 hectares.[5] Land use varies across

Table 12.1 Explanatory variables and expected relation with CSA implementation index

Dependent variable	
Index of implementation of CSA practices related to trees: Measures respondent's recognition of implementation of agroforestry/living fences/reforestation; tree diversity; and tree use diversity	
Explanatory variables	*Expected relation*
Sex	(+)
Age	(+)
Education	(+)
Access to information	(+)
Land control	(+)
Group participation	(+)
Existence of empowering gender norms (whether or not 1 or both primary decision makers promote empowering gender norms)	(+)
Labor availability (number of working age household members)	(+)
Tenure security (whether or not the household rents any land)	(-)
Off-farm income (existence of primary decision-makers who work off-farm)	(+)
Total farm size	(+)
Coffee land use	(+)
Basic grain land use	(-)
Home garden land use	(-)
Pasture land use	(-)

households; however, on average, households dedicate 17% of their land to coffee production; 65% of farmland to basic grain production; 2% of farmland to home gardens. Eighty-eight percent of households own at least 1 parcel for farming, while only 49% own all of their land, which implies that 51% of the households rent at least 1 parcel for farming. In about 62% of the households, women own at least one parcel individually or jointly with their spouse or other household member/s. Also, in just 14% of the households surveyed at least one primary decision maker had worked off-farm over the last year.

Table 12.3 presents summarized characteristics of the respondents by gender. In particular, the results highlight differences and similarities in dimensions of women's and men's empowerment. Agency in decision making demonstrates a

Table 12.2 Characteristics of respondents

Characteristics of respondents by gender	*Men (n = 239)*		*Women (n = 253)*	
Sex	48.5%		51.5%	
Age	18–78, M 46		18–89, M 43	
Education*				
None	39.8%		37.2%	
Primary	52.3%		47.8%	
Secondary	5.9%		11.5%	
Technical	0.0%		2.0%	
Higher	2.1%		1.6%	
Binary variables	No	Yes	No	Yes
Access to information*	51.1%	49.0%	60.6%	30.4%
Land control***	18.8%	81.2%	37.4%	62.6%
Group participation	82.4%	17.6%	80.3%	19.7%

*$p < 0.05$, **$p < 0.01$, ***$p < 0.001$

Table 12.3 Household characteristics

Household characteristics		
Total farm size (hectares)	.10–35.22, M 4.02	
% Coffee land use	0-.92, M .14	
% Basic grain land use	0–1, M .49	
% Home garden land use	0-.5, M .01	
% Pasture land use	0-.35, M .003	
Labor availability (number of working age household members)	0–8, M 3.3	
BINARY VARIABLES	No	Yes
Tenure security (whether or not the household rents any land)	48.9%	51.1%
Off-farm income (existence of primary decision makers who work off-farm)	74.0%	26.0%

significant gender equality gap, with 84% of men reporting participation in farm decision making in comparison to only 24% of women. Contrastingly, men and women show similar levels of access to information, with slightly more women (89%) than men (85%) reporting having received information related to the

CSA practices. Results also indicate a gender gap in land ownership, with 72% of men reporting ownership of at least one land parcel, in comparison to 62% of women. In general, group participation tends to be low among both women and men in the site. Slightly more women (21%) report group participation than men (18%).

Household implementation of CSA practices related to trees and women's empowerment

Table 12.4 presents the results of the probit model. In particular, the results suggest that when women primary decision makers are empowered, the household is more prone to implement CSA practices related to trees. Women's access to information and agency in decision making are both positively and significantly associated with household implementation of the CSA practices, at the 90% confidence level. Furthermore, women's land ownership is positively and significantly associated with CSA implementation related to trees, at the 95% confidence level, suggesting that households may be more prone to implement practices when women primary decision makers own land. The results indicate that women's group participation may be a particularly important factor for household implementation of CSA practices related to trees, showing that the variable is positively and significantly associated with implementation at the 99% confidence level. That access to information, agency in decision making and land ownership would not be as significantly associated with household CSA implementation, as group participation can be explained by the fact that information

Table 12.4 Results of linear regression model

Linear	*Coefficient*	*t stat*
Sex	.634***	8.48
Education	.198*	2.55
Access info	.334***	4.34
Land control	.407***	4.57
Group participation	.197*	2.02
1PD gender norms	.401***	4.02
2PD gender norms	.475***	4.64
Basic grains LU	−.214***	−5.61
Home garden LU	−.103**	−2.80
Basic grains^2	−.292***	−6.45
Constant	−.971***	−8.01
R^2 /adjusted R^2	.381/.364	

*$p < 0.05$, **$p < 0.01$, ***$p < 0.001$

on the CSA practices, seems to be largely accessible to both women and men (as demonstrated in Table 12.3). Also, although the results suggest a gender gap in land control, that the male primary decision maker alone owns land can often sufficiently enable household implementation of CSA practices. This may also explain why women's agency in decision making is not as significant; men's land ownership and decision-making power may often be enough to facilitate household implementation.[6] In contrast, both women and men may be equally limited from participating in groups, as mentioned previously, with women participating slightly more than men; the leadership skills and social capital that they are able to access via their participation may be particularly important for the household's capacity to implement CSA practices related to trees.

None of the land use variables show a significant association with the implementation of CSA practices related to trees, although all are negatively related. The negative relationship had been expected for all the types of land uses except coffee production. The lack of association could be due to the nature of the dependent variable, which measures whether the household implemented CSA practices related to trees, but does not assess the extent of implementation; consequently, the model might show a different degree of association with land use if a different type of dependent variable is used.

Notwithstanding, other household level variables show significant relationships with household implementation of CSA practices. For instance, as the household's cultivated farm size increases, so does the tendency for implementation of CSA practices related to trees, at the 95% confidence level. Land tenure is negatively and significantly associated with implementation at the 95% confidence level as well; this suggests that when a household rents at least one parcel of land, there is less tendency for implementation of CSA practices related to trees. Participation in off-farm work is positively and significantly associated with implementation at the 99% confidence level. The result suggests that the additional income source can be an important factor for implementation of CSA practices related to trees.

Qualitative results

While the qualitative information affirms several of the relationships identified through the quantitative analysis, they particularly help explain dynamics among the explanatory variables. For example, women's access to productive resources, decision-making power and group participation and normative structures concerning household gender roles can facilitate greater combined spousal learning and action on climate-smart practices. At the same time, implementation of CSA practices related to trees can significantly depend on household-level factors.

CSA *practices related to trees*

All men and women interviewed had knowledge of CSA practices related to trees on farms, although men may have demonstrated more knowledge learned via formal workshops and trainings in comparison to women. All the interviewees implemented CSA practices related to trees on farms to differing extents on their land as well; however, those couples involved in cocoa or coffee production demonstrated slightly more knowledge and active implementation of the practices. Wealth, represented conceptually by land size, also emerged as a possible significant factor for the implementation of CSA practices related to trees. The household with the smallest land size was limited from implementing practices like agroforestry and reforestation, despite that both the man and woman primary decision makers had knowledge of the practices, the male in particular having participated in several trainings on the themes.

Focusing on those households in which both women and men control land, results also highlight the influence of land use on implementation of agroforestry-related practices. Of the seven couples interviewed, all men were landowners, and three of the women owned land either jointly with or separately from their partners. In a household where the woman primary decision maker had sole ownership of a parcel, there was a difference in knowledge of CSA practices related to trees between the couple, with the woman demonstrating a greater breadth of knowledge than her partner. However, it must be highlighted that the male spouse dedicated the majority of his land to livestock production. Also, the woman primary decision maker did not implement the practices significantly on her parcel (although she did implement other types of CSA practices). The households in which women owned land jointly with their spouses demonstrated extensive implementation of CSA practices related to trees on farms; these were also coffee-producing households.

Empowerment Indicators

Women who owned land also tended to participate in groups and had access to information from external sources on CSA practices related to trees on farms. They had also accessed information on various additional issues related to climate change, food security and nutrition. Group participation – women-specific, with women and men, or in some cases both – was important means of accessing the information. Outside entities like NGOs or agricultural research centers had developed the groups. Correspondingly, the women accessed information through their participation in development projects with the NGOs or research centers, participating in trainings and workshops in the process. Women who did not own land did not access information through external sources on CSA practices related to trees; however, they did via their families, either their fathers or spouses or both.

The majority of men reported accessing information on CSA practices related to trees via external sources. Notwithstanding, in some cases of those households where women owned land as well, women's access to information from external sources served to facilitate men's and the household's greater knowledge and learning. For instance, in the household where the female primary decision maker had sole ownership of a land parcel, the spouse did not directly access information on CSA practices via external entities; however, he did access information from outside sources via his wife and the information she obtained through her work with NGOs and her participation in trainings on the subject. The female primary decision maker also implemented the information she had acquired via other family members including her eldest son, whom she would direct in work with her on her parcel. Similarly, in a household where the male and female primary decision makers owned land jointly, the male spouse shared in the information and capacity building that his partner acquired via external sources by often participating with her in the trainings and group meetings to which she was invited. The couples with joint land ownership had acquired the jointly held land through a program facilitated by a Nicaraguan NGO. The program had the goal of promoting women's land ownership through joint land titling with spouses.

The results suggest that norms that promote women's and men's shared participation in agricultural decision making can engender women's and men's enhanced access to information. The tendency for women's increased access to information and involvement in decision making may be associated with women's land control. Most men supported women's participation in groups, in capacity building and in development projects organized by outside entities. Men respondents' reasons for supporting women's group participation ranged from concern for securing women's human rights to more practical interest based in the belief that both spouses are obliged to share in the household's agricultural work, thus necessitating that both women and men learn about agriculture. Other reasons were grounded in the idea that both women and men have equal capacities to make decisions for the household; consequently, women's capacity building benefits the household. Furthermore, as mentioned in the previous discussion on access to information, one male primary decision maker recognized the benefit of his wife's participation in groups and trainings because they participate together and he shares in the capacity building. In general, those couples who highlighted the importance of shared, consultative decision-making processes were those who had decided to participate in development projects together, like the joint land titling program; consequently, these were also the households in which women and men owned land jointly.

Discussion and conclusions

Together, the quantitative and qualitative results indicate that attention to aspects of women's empowerment can be important for promoting increased implementation of CSA practices related to trees on farms. Various household level and land-use variables are critical as well. For women household primary decision makers in the site in Nicaragua, group participation can contribute to their households' implementation of CSA practices like agroforestry, living fences and reforestation. Though not as significant, women's land ownership, agency in decision making and access to information on the CSA practices can also be important factors for households' increased implementation. While indigenous knowledge on the use of trees on farms is important, technical information on CSA practices can complement this knowledge and contribute to household confidence in agricultural decision making and agency in farm management. Normative structures concerning household gender roles can influence women's involvement in agricultural decision making and access to and control of productive resources like capacity building, leadership skills and farmland. It must be highlighted that increased farm size and extra sources of capital (through off-farm labor) can promote household implementation of CSA practices related to trees, while insecure land ownership can deter it as well. Land use may prove more significant in cases in which extent of implementation is assessed; for instance, it is possible that coffee-producing households would engage in agroforestry practices more than those who dedicate significant portions of their farmland to home gardens or basic grain production. That the dimensions of women's empowerment can be critical for household implementation of the CSA practices is indicative of the significance of gender inequalities in access to productive resources and agency. While commonly disadvantaged in these areas in comparison to men, women's empowerment can contribute to enhanced capacities of the household to make informed decisions for farm management and improved resilience.

The study is particularly important, given the limited amount of research on the contributions of household members' behaviors to implementation of agricultural innovation and technology. While most adoption studies tend to target the household head, our analysis focuses on primary decision makers. Consequently, the analysis of the association of women's empowerment indicators with CSA implementation is not limited to women household heads but rather takes into account women living in male-headed households as well.

Critically, the results indicate the importance of interventions that incorporate processes to change gender inequalities and promote women's empowerment in order to scale out implementation of CSA practices related to trees. This can require engagement with local norms and institutions and concerted efforts to consider various aspects of women's empowerment. For instance, enhancing women's group participation can necessitate working with community groups, civil society organizations and producer organizations to promote group processes that are equally accessible to women and men. In some cases, this might require development and support of new or existing women-specific groups (Pandolfelli et al., 2008). Coordination with government agencies to support extension services that reach and benefit both women and men and with other local organizations to provide trainings and technical assistance relevant to CSA can be necessary as well.

Similarly, initiatives to enhance women's household decision-making power will depend critically on challenging local norms and institutions and coordination with diverse partners. Past examples have included trainings for couples on gender-equal relations in the household, like the Gender Action Learning Systems methodology promoted by Fair Trade organization TWIN in Malawi (Farnworth et al., 2018) and the Hanns R. Neumann Stiftung gender program targeting smallholder coffee producers in western Uganda (Lecoutere and Jassogne, 2016). Also, programs on joint land titling between spouses, like that facilitated through the NGO in the site in Nicaragua and like India's land reallocation and titling program in West Bengal (Santos et al., 2013), can tend to promote joint agricultural decision making. In this way, through collaboration with initiatives that address key gender equality gaps, CSA interventions can not only achieve impacts at scale but also promote gender transformation.

Acknowledgments

The authors gratefully acknowledge the Independent Science and Partner Council (ISPC) of the CGIAR for its support of the research project. The authors would like to give a special thanks to Geovana Carreno for initial comments on the model design as well. This research was carried out by CIAT as part of the CGIAR Research Program on Forests, Trees and Agroforestry (CRP-FTA). This collaborative program aims to enhance the management and use of forests, agroforestry and tree genetic resources across the landscape from forests to farms. CIFOR leads CRP-FTA in partnership with Bioversity International, CATIE, CIRAD and the World Agroforestry Centre. This work was also implemented as part of the CGIAR Research Program on Climate Change, Agriculture and Food Security (CCAFS), which is carried out with support from the CGIAR Trust Fund and through bilateral funding agreements. For details, please visit https://ccafs.cgiar.org/donors. The views expressed in this document cannot be taken to reflect the official opinions of these organizations.

Notes

1 We do not analyze measurements of time allocation; also, as information from the survey related to the remaining WEIA domain, income control, is extensive, this will be the subject of analysis for future papers.
2 https://gender.cgiar.org/themes/gennovate/
3 The question on source of information about the practice asks if the respondent received any information in the last 12 months. It is possible that the respondent might have received information in the past, earlier than previous 12 months, and therefore considers it prior knowledge.
4 The agroforestry adoption studies highlight access to extension as an important influence; however, this is accounted for by the abovementioned variable on access to information.
5 A household farm is the combination of all the land parcels owned or rented and cultivated by the household.
6 Intra-household land control trends can differ across regions. For example, among smallholder households in Africa, women commonly tend to have their own plots of land, separate from male household members.

References

Alkire, S., Meinzen-Dick, R., Peterman, A., Quisumbing, A.R., Seymour, G. and Vaz, A. (2013). *The women's empowerment in agriculture index*. OPHI Working Paper 58. University of Oxford.

Amores Contreras, F.M. (2015). *Contribución de los árboles en finca a los medios de vida de familias rurales en dos sitios contrastantes de Nicaragua*. [MS Thesis]. Turrialba, Costa Rica: Centro Agronómico Tropical de Investigación y Enseñanza.

Asher, K. and Sijapati Basnett, B. (2016). Assessment: Gender equality as an entitlement: An assessment of the UN women's report on gender equality and sustainable development 2014. *Development and Change*, 47(4), pp. 952–964.

Ashraf, J., Pandey, R., de Jong, W. and Nagar, B. (2015). Factors influencing farmers' decisions to plant trees on their farms in Uttar Pradesh, India. *Small-scale Forestry*, 14, pp. 301–313.

Badstue, L., Kantor, P., Prain, G., Ashby, J. and Petesch, P. (2014). Innovation and development through transformation of gender norms in agriculture and natural resource management: A global comparative research initiative. *Concept Note*. Available at: https://gender.cgiar.org/wp-content/uploads/2015/09/CN-Global-Study-Gender-NormsAgency_modified-Jan-2015.pdf [Accessed 17 November 2017].

Bee, B. and Sijapati Basnett, B. (2016). Engendering social and environmental safeguards in REDD+: Lessons from feminist and development research. *Third World Quarterly*, 38(4), pp. 787–804.

Carr, E.R. (ed.) (2014). *Assessing Mali's Direction Nacionale de la Météorologie agrometeorological advisory program: Preliminary report on the climate science and farmer use of advisories*. Washington, DC: USAID.

Carr, E.R. and Onzere, S.N. (2017). Really effective (for 15% of the men): Lessons in understanding and addressing user needs in climate services from Mali. *Climate Risk Management*, pp. 1–14.

Carr, E.R., Onzere, S.N., Kalala, T., Rosko, H.M. and Davis, J. (2016). *USAID/Mali climate change adaptation activity (MCCAA) behavioral baseline survey: Final synthesis report.* Washington, DC: USAID.

Carr, E.R. and Owusu-Daaku, K.N. (2016). The shifting epistemologies of vulnerability in climate services for development: The case of Mali's agrometeorological advisory programme. *Area*, 48(1), pp. 7–17.

Catacutan, D. and Naz, F. (2015). Gender roles, decision-making and challenges to agroforestry adoption in Northwest Vietnam. *International Forestry Review*, 17(S4), pp. 22–32.

Chi, T.T.N., Paris, T., Anh, T.T.T., Duy, L. and Loan, D.T. (2015). *Enhancing the roles of women in rice farming as an adaptation strategy to climate change risks: A case study in submergence villages in Hau Giang province, South Vietnam.* Hanoi: Cuu Long Rice Research Institute.

Conteh, A.M.H., Moiwo, J.P. and Yan, X. (2016). Using a logistic regression model to analyze alley farming adoption factors in Sierra Leone. *Small-scale Forestry*, 15, pp. 109–125.

Cornwall, A. (2016). Women's empowerment: What works? *Journal of International Development*, 28, pp. 342–359.

Coulier, M. (2016). *ACIS project baseline survey report – Vietnam.* Hanoi: CCAFS-ICRAF-CARE.

da Cunha, E.M. (2006). *Mutirão e trabalhadoras de igarapé-miri: açaí como alternativa econômica no contexto de gênero [Mutirão and workers from Igarapé-Mirim: açaí as economic alternative in gender context]*. Belém (Brazil): Núcleo de Altos Estudos Amazônicos. *Papers do NAEA Report* No.: 206. Portuguese.

Dhakal, A., Cockfield, G. and Narayan Maraseni, T. (2015). Deriving an index of adoption rate and assessing factors affecting adoption of an agroforestry-based farming system in Dhanusha District, Nepal. *Agroforestry Systems*, 89, pp. 645–661.

Doss, C. (2017). Including both equity and efficiency claims for international development. *Canadian Journal of Development Studies*, 38(4), pp. 553–557.

Elias, M. and Arora-Jonsson, S. (2016). Negotiating across difference: Gendered exclusions and cooperation in the shea value chain. *Environment and Planning D: Society and Space*, 0(0), pp. 1–19.

Evans, K., Flores, S., Larson, A.M., Marchena, R., Müller, P. and Pikitle, A. (2016). Challenges for women's participation in communal forests: Experience from Nicaragua's indigenous territories. *Women's Studies International Forum*, pp. 1–10.

FAO. (2011). *The state of food and agriculture. Women in agriculture: Closing the gender gap for development*. Rome, Italy: FAO.

Farnworth, C., Stirling, C., Chinyophiro, A., Namakhoma, A. and Morahan, R. (2018). Exploring the potential of household methodologies to strengthen gender equality and improve smallholder livelihoods: Research in Malawi in maize-based systems. *Journal of Arid Environments*, 149, pp. 53–61.

Higuchi, A., Masahiro, M. and Fukuda, S. (2012). The impact of socio-economic characteristics on coffee farmers' marketing channel choice: Evidence from Villa Rica, Peru. *Sustainable Agriculture Research SAR*, 1(1).

Jazíbi Cárcamo, N., Vásquez García, V., Zapata Martelo, E. and Nazar Beutelspacher, A. (2010). Género, trabajo y organización. Mujeres cafetaleras de la Unión de productores orgánicos San Isidro Siltepec, Chiapas. *Estudios Sociales*, 18(36), pp. 156–176.

Jost, C., Kyazze, F., Naab, J., Neelormi, S., Kinyangi, J., Zougmore, R., Aggarwal, P., Bhatta, G., Chaudhury, M., Tapio-Bistrom, M., Nelson, S. and Kristjanson, P. (2016). Understanding gender dimensions of agriculture and climate change in smallholder farming communities. *Climate and Development*, 8(2), pp. 1–12.

Kabeer, N. (1999). Resources, agency, achievements: Reflections on the measurement of women's empowerment. *Development and Change*, 30, pp. 435–464.

Kabeer, N. and Subrahmanian, R. (1996). *Institutions, relations and outcomes: Frameworks and tools for gender-aware planning*. IDS Discussion Paper 357. Brighton: Institute of Development Studies.

Kantor, P., Morgan, M. and Choudhury, A. (2015). Amplifying outcomes by addressing inequality: The role of gender-transformative approaches in agricultural research for development. *Gender, Technology and Development*, 19(3), pp. 292–319.

Koczberski, G. (2007). Loose fruit mamas: Creating incentives for smallholder women in oil palm production in Papua New Guinea. *World Development*, 35(7), pp. 1172–1185.

Kristjanson, P., Bryan, E., Bernier, Q., Twyman, J., Meinzen-Dick, R., Kieran, C., Ringler, C., Jost, C. and Doss, C. (2017). Addressing gender in agricultural research for development in the face of a changing climate: Where are we and where should we be going? *International Journal of Agricultural Sustainability*, 15(5), pp. 482–500.

Lecoutere, E. and Jassogne, L. (2016). *"We're in this together": Changing intra-household decision making for more cooperative smallholder farming*. Working Paper. Institute of Development Policy and Management.

Meijer, S.S., Sileshi, G.W., Kundhlande, G., Catacutan, D. and Nieuwenhuis, M. (2015). The role of gender and kinship structure in household decision-making for agriculture and tree planting in Malawi. *Journal of Gender, Agriculture and Food Security*, 1(1), pp. 54–76.

Mello, D. (2014). *Collective microenterprises and rural women's economic empowerment in Brazilian Amazonia*. [PhD thesis]. Gainesville, FL: University of Florida.

Pandolfelli, L., Meinzen-Dick, R. and Dohrn, S. (2008). Gender and collective action: Motivations, effectiveness and impact. *Journal of International Development: The Journal of the Development Studies Association*, 20(1), pp. 1–11.

Pandolfelli, L., Meinzen-Dick, R. and Dohrn, S. (2007). *Gender and collective action: A conceptual framework for analysis*. Collective Action Working Paper 64. Washington, DC: International Food Policy Research Institute (IFPRI).

Poulsen, E., Sakho, M., McKune, S., Russo, S. and Ndiaye, O. (2015). *Exploring synergies between health and climate services: Assessing the feasibility of providing climate information to women farmers through health posts in Kaffrine, Senegal*. CCAFS Working Paper no. 131. Wageningen, Netherlands: CGIAR Research Program on Climate Change, Agriculture and Food Security (CCAFS). Available at: https://cgspace.cgiar.org/bitstream/handle/10568/68141/CCAFS%20WP%20131.pdf?sequence=1.

Roncoli, C., Jost, C., Kirshen, P., Sanon, M., Ingram, K.T., Woodin, M., Somé, L., Ouattara, F., Sanfo, B.J., Sia, C., Yaka, P. and Hoogenboom, G. (2009). From accessing to assessing forecasts: An end-to-end study of participatory climate forecast dissemination in Burkina Faso (West Africa). *Climatic Change*, 92(3–4), pp. 433–460.

Santos, F., Fletschner, D., Savath, V. and Peterman, A. (2013). *Can government-allocated land contribute to food security? Intrahousehold analysis of West Bengal's microplot allocation program*. IFPRI Discussion Paper 01310.

Serra, R. and McKune, S. (2016). *Climate information services and behavioral change: The case of Senegal*. Sahel Research Group Working Paper No. 010. University of Florida. Available at: https://sahelresearch.africa.ufl.edu/files/SRG_WorkingPaper_Serra_McKune_Final.pdf.

Shreckenberg, K. and Marshall, E. (2006). Women and NTFPs: Improving income and status? In: E. Marshall, K. Schreckenberg and A.C. Newton, eds., *Commercialization of non-timber forest products: Factors influencing success. Lessons learned from Mexico and Bolivia and policy implications for decision-makers*. Cambridge: UNEP World Conservation Monitoring Centre, pp. 77–80.

Twin. (2013). *Empowering women farmers in agricultural value chains*. Rep. London: Twin.

Venkatasubramanian, K., Tall, A., Hansen, J. and Aggarwal, P.K. (2014). *Assessment of India's integrated agro-meteorological advisory service program from a farmer perspective*. CCAFS Working Paper no. 54. Wageningen, Netherlands: CGIAR Research Program on Climate Change, Agriculture and Food Security (CCAFS). Available at: https://cgspace.cgiar.org/rest/bitstreams/34467/retrieve.

Versteeg, S., Pilegaard Hansen, C. and Pouliot, M. (2017). Factors influencing smallholder commercial tree planting in Isabel Province, the Solomon Islands. *Agroforestry Systems*, 91, pp. 375–392.

13 Kinship structures, gender and groundnut productivity in Malawi

Edward Bikketi, Esther Njuguna-Mungai, Leif Jensen and Edna Johnny

Introduction

Kinship structures in rural African communities

Social structures in rural Sub-Saharan Africa (SSA) are mainly characterized by two types of kinship and descent – matrilineal and patrilineal. Kinship refers to the networks that connect individuals as relatives, while descent constitutes the social rules that identify individuals with a selected category of their kin for specific purposes (Kaarhus, 2010). Individuals form a lineage when they trace descent from a common ancestor. Matrilineal descent is traced through women (mothers, daughters etc.), while patrilineal is traced through men (fathers, sons etc.) (Kaarhus, 2010). In a matrilineal system, a married couple normally takes up residence in the wife's village (uxorilocal residence), while in a patrilineal system, the married couple takes up residence in husband's village (virilocal residence) (Meijer et al., 2015). In most patrilineal communities, women are not allowed to inherit, own and/or control land and most assets, leaving them wholly dependent on their husbands or male relatives for their livelihoods, financial security and well-being (Lastarria-Cornhiel, 2006; Doss et al., 2013; Bikketi et al., 2016). In matrilineal communities, women inherit, own and control the use of land and other assets. As such, lineage systems, operating through cultural norms and institutions, have profound implications for gender differences in household headship; access, management and ownership of household resources; opportunity structures; and degree of agency among actors (World Development Report, 2012, pp. 72–91; Doss, 2017).

Focus and objective of the study

The study focused on groundnut productivity among farmers from matrilineal and patrilineal communities in Malawi. In the case of matrilineal communities, the issue is whether women are empowered, enjoy full rights and exercise decision making and control of productive resources located in their domain. If so, then it is hypothesized that women will be as productive as men if not more. The overall objective was to assess the effects of kinship structures and lineage systems on groundnut production and productivity in Malawi. The chapter has six sections. Section 1 is the background, section 2 is the conceptual framework, section 3 is the research methodology, the research findings are presented in section 4, the discussion of the findings is presented in section 5, while conclusion and policy implications are presented in section 6.

Conceptual framework

The conceptual framework (presented in Figure 13.1), is a multilayered conceptualization of factors that that affect, (directly or indirectly), groundnut production in rural Malawian villages. However, the framework is not exhaustive of all factors. Each of the four layers is discussed below.

Kinship structure

The outer layer is the kinship structure that allocates certain rights and privileges to men or women. In Malawi, matrilineal and patrilineal structures have institutions associated to ownership, inheritance and residence. In matrilineal-matrilocal societies, land and land rights are transferred through women (Kaarhus, 2010; Meijer et al., 2015). The husband pays "bride service" by working for a negotiated

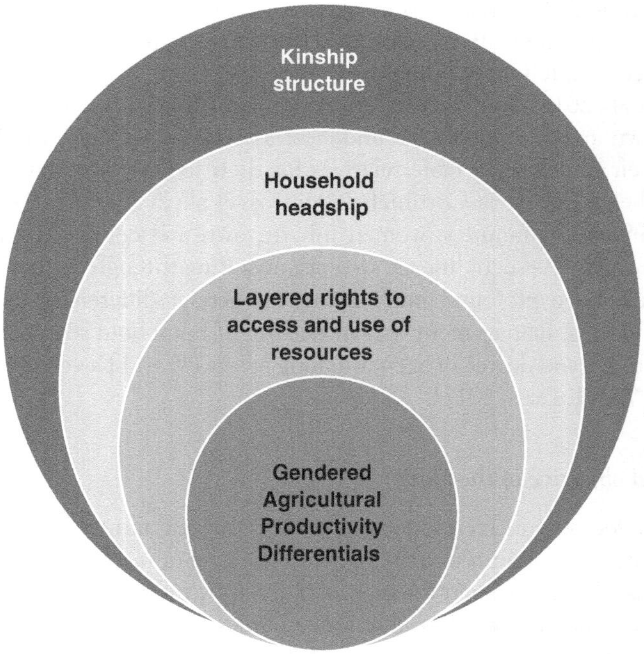

Figure 13.1 Conceptual framework

duration and has no control over land rights. In event of divorce or death of the wife, the husband loses the user rights over the land and is expected to return to his original village, leaving the children with the wife or her family, as children belong to the matrilineal kin (Meijer et al., 2015). In patrilineal-patrilocal societies, land and land rights are transferred through men, and upon divorce, the woman returns to her original village, while the children remain in the husband's village (Kaarhus, 2010). Widows may sometimes remain in the husband's village if bride wealth was paid and if the relatives of the husband permit (Kaarhus, 2010; Meijer et al., 2015).

Household headship

The concept of household headship is determined by the kinship structures, and based on this, the study categorized households into three types: male-headed households (MHH), female-headed households (FHH) and female-managed households (FMH). Female-managed households are those in which male spouses have out-migrated in search of off-farm livelihoods within Malawi or elsewhere in Southern Africa.

Layered rights to access and use of resources

This layer encompasses the rights, privileges and opportunity structures that prevail for the different male and female farmers, based on the outcomes of the intersecting kinship structures and the household headship arrangements. The concept involves five rights: (1) access to resources, (2) withdrawal-ability to use the output from the resources, (3) management rights or decision making about the use of resources (including crop choice and agricultural technologies), (4) exclusion rights and (5) rights of alienation, or decision making about the sale, lease, transfer or bequeathing of the resource (Rao, 2016). The plot manager needs to have at least the first three rights mentioned above. On this basis, the concept of plot manager intersects with gender to categorize the plot managers into three typologies: male manager, female manager and jointly managed (cf. Slavchevska et al., 2017).

Gendered agricultural productivity differentials

The last layer is a representation of the gendered groundnut productivity attained at the plot level, which is a manifestation of all the interacting layers.

Research methodology

Research sites

Malawi is a landlocked country in southern Africa bordered by the United Republic of Tanzania to the north and northeast, Mozambique to the east, south and southwest and Zambia to the west (FAO, 2015). Malawi is among the world's least developed, with 90% of its population living in rural areas and a predominantly agrarian economy in which agriculture accounts for 27% of its GDP and 85% of export revenues (Takuji et al., 2016). Groundnut is a major legume widely grown by Malawi's smallholders, who account for 93% of total production. It provides family nutrition, incomes, soil fertility and livestock feed supplements (Takuji et al., 2016).

Fieldwork was conducted in three districts (Mchinji, Mzimba and Mangochi), purposively selected for groundnut production and having different kinship structures. As shown in Figure 13.2, Mchinji district is in central Malawi. It covers an area of 3,356 km^2 and had a population of about 630,135 in 2017. The population is mainly from the Achewa ethnic Bantu group. Chichewa is the main language and is also the national language of Malawi. Mzimba district is in northern Malawi. It has an area of 10,430 km2. It is the largest district in Malawi, with an estimated 2017 population of 1,034,967. The inhabitants are mainly from the Tumbuka and Ngoni ethnic groups; the latter hailed from South Africa. They

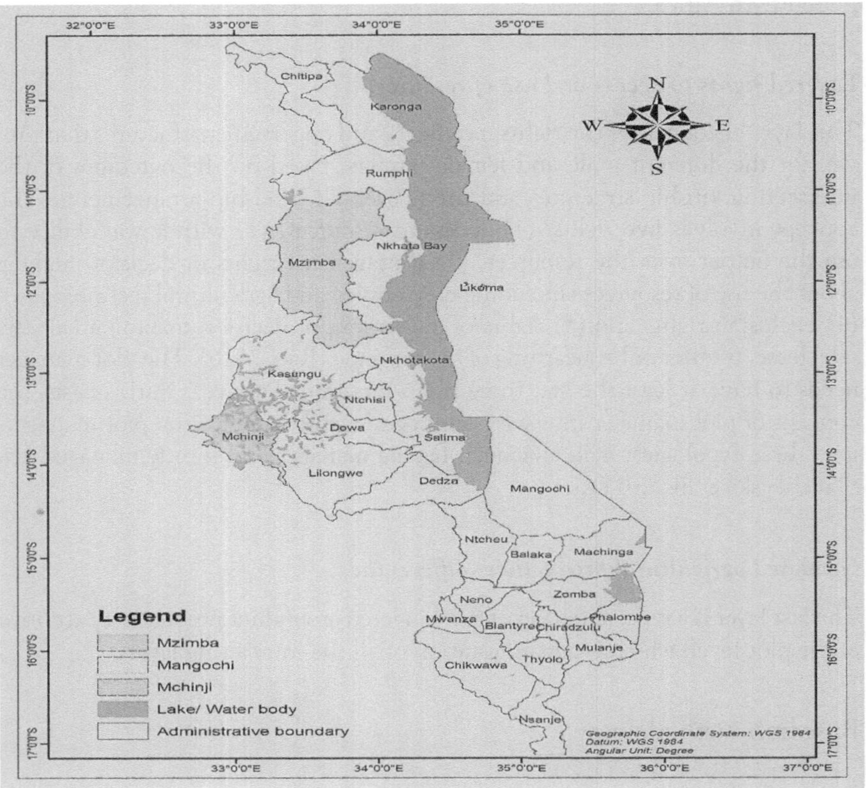

Figure 13.2 Map of Malawi with study areas.

Source: authors' design

speak Chitumbuka. Mangochi is in southern Malawi and has a population of 1,066,914 population covering an area of 6,273 km2. The district is on the floodplain for Lake Malawi. The main inhabitants are from the Ayao ethnic group, and the main language is Chiyao.

Exploratory survey

For the exploratory survey, Extension Planning Areas (EPAs) were purposively sampled based on groundnut producing populations and kinship structures. Within each selected EPA villages were purposively sampled based on categorization of household headship; MHH, FHH and FMH. In each village respondents were purposively sampled based on gender, kinship and management plot levels. Forty farmers were interviewed from five focus group discussions (eight farmers per group, three female farmer groups and two male farmer groups). Five other key informant interviews were undertaken for extension staff and managers of National Association of Smallholders in Malawi (NASFAM) and four case histories (two each for the matrilineal and patrilineal). Table 13.1 presents the EPAs, villages household headship and kinship structures sampled for each district. This information was used to inform the development of the structured survey.

Structured survey

For the structured survey, random and snowball sampling were used and involved a total of 285 respondents (Mchinji 100, Mzimba 113 and Mangochi 72). The survey was used to collect data on sociodemographic profiles; ownership of assets and resources; access to inputs and advisory services; groundnut productivity at the household and plot levels. While the sample size was not representative of the districts' population, it enabled in-depth insights into kinship structures and productivity of the different contexts.

Data collection and analysis

The study used a mixed-methods design to collect sex-disaggregated data. The combined use of qualitative and quantitative collection generally provides a

Table 13.1 Districts, extension planning areas (EPAS) and villages and households sampled

District	*Extension Planning Area*	*Villages*	*Household Headship*	*Kinship Structures*
Mchinji	Kalulu, Mikundi	31	FHH	Matrilineal
Mzimba	Emfeni, Embegueni, Mbawa, Kapalankwali	40	MHH	Patrilineal
Mangochi	Mtiya, Namwera	25	FMH	Matrilineal

Source: Survey Data.

better empirical understanding than using each approach alone (Terrell, 2012). This integration fostered triangulation and more comprehensive evidence to address the study objectives. Content analysis and ATLAS Ti software were used to analyze qualitative data while STATA, MS Excel and the Statistical Package for Social Sciences (SPSS) were used to analyze quantitative data. The qualitative analysis used a context-sensitive approach that considers social, historical and temporal experiences in space and time to collate, code and transcribe the data. This enabled careful comparative case analyses and extrapolation of meaningful themes and patterns through reflexivity and iteration. Quantitative analysis involved descriptive statistics and OLS regression to estimate bivariate and multivariate relationships between independent variables and groundnut productivity.

4. Research findings

Respondent profiles

Table 13.2 presents the respondents profiles in the study areas. The average household sizes are between four and five persons, and most respondents had attained primary school education. The average ages indicate that Mzimba has more elderly farmers, and overall very few youths are engaged in groundnuts production in the three districts, reflecting an untapped youth dividend (cf. OECD Development Center, 2018). Land sizes are less than a half hectare and are declining due to inheritance and fragmentation as the population increases (Lovo, 2016). Land is mostly family or lineage property that has transitioned from typical customary tenure held in trust by chiefs for villagers (cf. Peters, 2010).

Table 13.2 Respondents' profiles

Districts	*Mangochi (FMH)*	*Mchinji (FHH)*	*Mzimba (MHH)*
Respondents			
No. of villages	25	31	40
No. of respondents	72	100	113
Lineage system	Matrilineal	Matrilineal	Patrilineal
Gender	Male: 0% Female: 100%	Male: 51% Female: 49%	Male: 52% Female: 48%
Households	73	51	57
Ethnic tribes	Yao	Chewa	Tumbuka
Average age	37.1 years	45.5 years	47.2 years
Average household sizes	4 people	5 people	5 people
Educational Attainment			
Average no. of years schooling	5.97	13.31	9.52
Land sizes			
Average land sizes (ha)	0.342	0.404	0.270

Source: survey data.

Culture, gender relations and livelihoods

Culturally, the matrilineal structure has inherently prevailed in 20 districts of central and southern Malawi despite being subjected to change by contemporary forces (Mwambene, 2005), while patrilineal structure prevails in all the six districts of northern Malawi but one in the south (Nsanje). These forces included slave traders (Arab and Yao), British and Portuguese colonialists, Scottish Missionaries, and incursions by the ferocious warriors of the Ngoni tribe from southern Africa. Post-independence governments have reinforced ideologies of the colonialists about matrilinity as detrimental to development (Mwambene, 2005; Takane, 2008; Kaarhus, 2010; Jennifer, 2014).

The Chewa and Yao ethnic tribes practice matrilineal-matrilocal kinship. These are described briefly. The Chewa livelihoods revolve around subsistence agriculture, producing maize, vegetables, beans, groundnuts and soybeans. Groundnuts and soybeans have gained prominence as cash crops. Tobacco used to be the main cash crop, but farmers are switching to soybeans due to declining tobacco prices.

Chewa culture has assimilated practices from patrilineal kinship and Christianity that colluded to recast gendered roles of women to fit into male-headed nuclear patrilineal units, compounded by global monetary economy, that have transformed households' livelihoods, productivity patterns and division of labor to the disadvantage of women (Peters, 2010; Jennifer, 2014). However, marriage is still regularized under matrilineal-matrilocal customs, and the post-marital residence is uxorilocal with the woman as the household head (FHH), the man pays bride service locally known as "Chikamwini" for a negotiated length of time (cf. Dean, 2015). As confirmed by respondents in the case histories and ethnography:

> The husband pays "bride service" by working for a negotiated amount of time, services include; cultivation of land, siring children to keep his wife's lineage among others, but he has no control over his wife's land rights.
>
> (Personal communication of case-histories and ethnography respondents in Mitanda village, Kalulu EPA and Kalonga village Mikundi EPA, Mchinji district)

Land is inherently under family ownership and governed by customary law. Respective families along with village headmen and headwomen facilitate land inheritance and arbitrations. Landowners are accorded higher status, however; owning more means increased labor which is a major constraint for women. The respondents of the male and female FGD mentioned that:

> One of our traditional ways of managing labour shortage is through "ganyu," a social arrangement where men and women work for "food" normally maize as a mode of payment after completing a task in cases of widows, the elderly farmers, sick and the needy it is considered as benevolence.
>
> (Personal communication of respondents of the female FGD in Sendeza village, Kalulu EPA, Mchinji district)

> Ganyu labour is essential in building and conserving social networks in the villages, and it creates strong relations with each other.
>
> (Personal communication of respondents of the Male FGD in Matutu village Mikundi EPA Mchinji district)

Among the Yao, livelihoods revolve around fishing, tourism and subsistence farming of maize, rice and other food crops, with groundnuts as cash crops. Most of the Yao people don't keep livestock except a few goats, pigs and chickens. Many of those living near Lake Malawi depend on fishing to supplement their diets and incomes.

The Yao are predominantly Muslim dating back to the 18th century. According to Matinga (2015), the Yao accepted the expansion of Islam in Malawi as a way of modernism and were exposed to it through Arab traders. Like the Chewa, marriage and residence are matrilineal-matrilocal; however, divorce and separation are common, given the excessive out-migration of men, resulting in female-managed households (FMH). Men are constantly trying to gain status and thus mostly seek redress through negotiated bride service time, out-migration and/or divorce. As alluded to by respondents through the case history and mini-ethnography:

> The husband is more of a dependent worker in the wife's family, most of these marriages don't last, we have a lot of divorced and separated women in Mangochi as their husbands have left for South Africa and some never return because of the cultural obligations of the Chikamwini.
>
> Case history and mini-ethnography Mbalowa village, Namwera EPA Mangochi district)

These marriages are termed "walking marriages" due to their unstable nature (Jennifer, 2014). Land and land rights are inherited through the female lineage (cf. Mwambene, 2005; Takane, 2008; Matinga, 2015). Men are generally responsible for fishing, agriculture, and most trading activities; women are responsible for tasks related to maintenance and running of the household, among other reproductive roles.

In both Chewa and Yao, in the event of divorce or death of the wife, the husband loses user rights over his wife's land and is expected to return to his original village, leaving the children with the wife or her family, as children belong to the matrilineal kin (Meijer et al., 2015). Through the case histories and mini-ethnography, the respondents pointed out that:

> Our villages consist of groups of relatives we call "mafuko," these are our sons-in-law known as "akamwini," and daughters-in-law known as "akamwana" their children and other people related to the rest of the villagers by blood or marriage. However, in cases of divorce the akamwini leaves with only his clothes and a blanket just like he came into the village.
>
> (Personal communication of case-history and FGD respondents in Mitanda village, Kalulu EPA and Chibande village, Mikundi EPA, Mchinji district)

An extended group of families within the village is referred to as a "fuko," consisting of blood relations tracing descent through a female line to a first or common ancestress. Blood relations include adults, children and children born out of wedlock. From the case histories and ethnographies, respondents said:

> Each fuko is subject to the control of a matriarch locally referred to as "Kholo," The functions and obligations of a "Kholo" are, overseeing the welfare of the fuko and maintaining social order: (1) settling internal disputes; (2) providing general care and attention over all the members of the fuko; (3) intervening for compensation for wrongs done to any member of the fuko by an outsider; and (4) obtaining land for use of members of the "fuko," and to exercise general control over such land.
>
> (Personal communication of case-history and mini-ethnography respondents in Mitanda village, Kalulu EPA and Kalonga village Mikundi EPA, Mchinji district)

After the fuko there is "mbumba," a sorority group of sisters. These sorority groups are constituted in relation to a man, often the eldest brother of the sisters. Each mbumba is subject to the control of its oldest male member, locally called "Nkhoswe," who is usually, the sisters' eldest brother. The functions and obligations of the Nkhoswe are like those of Kholo, although the Nkhoswe must consult the Kholo in important matters like land, marriage and divorce. Lastly, there is "Banja." This social group is the basic household and is composed of a man and his wife, and their children.

> The "Banjas" are under the care of the husband. As the provider of the family, the husband is responsible for providing food and shelter for members. If a man has more than one wife, then he, together with each of his wives and their respective children, constitute a separate household.
>
> (Personal communication of case-history respondent in Luwemba village, Ntiya EPA, Mangochi)

Even though the husband has some control over his family, the general welfare of the Banja is in the hands of the Nkhoswe (Mwambene, 2005). In the absence of the wife's brothers, an elder sister or some other senior matrilineal relative acts as the Nkhoswe over their mbumba (cf. Takane, 2008). As mentioned by the respondents of the case histories and ethnography, the responsibility of the Nkhoswe involves:

> The Nkhoswe organises and oversees marriages protects the mbumba have access to land and other resources; and overseeing succession and inheritance rights. He is also responsible for the general welfare of his sisters and their children.
>
> (Personal communication of case-histories and ethnography respondents in Majiputa village, Ntiya EPA, Mangochi)

> In Chikamwini, fathers/husbands play a small role in these responsibilities. They have no authority in most decision-making, sometimes including control of what they farm produce, as the Nkhoswe allocates these farms to them. As a man, a husband's status is achieved through his mbumba.
>
> (Personal communication of case-history ethnography respondents in Mitanda village, Kalulu EPA and Kalonga village Mikundi EPA, Mchinji district)

In these matrilineal-matrilocal households there is a lot of interference by brothers and uncles of the women as the "custodians/guardians" of matrilineal groups locally referred to as "*Nkhoswe*" (cf. Jennifer, 2014). Consequently, men concentrate on agricultural enterprises such as groundnuts that will not attract intrusion, as they are considered women's crops and are attributed to the "*power over*" the male farmer by the Nkhoswe, as was confirmed by male respondents in the FGD conducted in Mikundi, Mchinji district;

> The Nkhoswe have a lot of influence in our homes and farming enterprises, as such mostly we men in Chikamwini choose to farm crops that do not attract attention and meddling such as groundnuts which are still considered to be women's crops.
>
> (Personal communication of FGD respondents in Matutu village, Mikundi EPA, Mchinji district)

The Tumbuka ethnic tribe practices patrilineal-patrilocal kinship. They were previously matrilineal, but because of incursion by the Ngoni and colonialists, they were encultured by the Ngoni patrilineal culture and Christianity, which also altered their social organization (Mushibwe, 2009; Peters, 2012). The Tumbuka accepted Ngoni culture as a means of identifying themselves with new rulers including paying bride wealth locally known as 'lobola' according to the Ngoni. The arrival of the missionaries reinforced the notion of bride wealth and replaced it with the term "dowry" (Mushibwe, 2009). The Tumbuka livelihoods also revolve around subsistence farming of food crops, supplemented by herding of goats and cattle, while tobacco, soybean and groundnuts are grown as cash crops. Respondents of the case histories confirmed that:

> The Tumbuka culture currently can be characterised as traditionally, inherently, matrilineal influenced by the Ngoni tribe with patriarchal practices, values and cultural traditions which do not recognize women as equal partners with men, for example polygamous marriages practiced by the Ngoni are currently accepted by the Tumbuka. We even refer to ourselves as Tumbuka-Ngoni.
>
> (Personal communication of case-history respondents in Mateyu village, Mbawa EPA, Mzimba district)

In Tumbuka culture, marriage, inheritance and residence remain patrilineal, with the man as the household head (MHH). Land and land rights are inherited through the male lineage by a generation of brothers in order of seniority. As confirmed by women in the FGD in Emfeni Mzimba District, Tumbuka women are not permitted to inherit land, a crucial primary resource for agricultural production:

> In our culture land is inherited by men, our access to land is through our fathers, brothers and husbands we can only use the land for agriculture but mostly we are not allowed to own land.
>
> (Personal communication of Female FGD Respondents in Chiwandauka-Chisi village, Emfeni EPA)

Upon demise of their husbands culturally, women are entitled to select successors from their husbands' generational agnates, who include brothers. However, due to Christianity and modern socioeconomic influences, most women currently select their eldest son and thus come under his authority. In Tumbuka culture, upon divorce, the wife returns to her original village while the children remain in the husband's village (Kaarhus, 2010; Meijer et al., 2015).

Further, Tumbuka women's experiences and beliefs reflect sociocultural traditional norms that tend to limit gender equality and compel women to accept and justify male hegemony at the expense of their own status and regard consequent inequalities as normal. Evidence demonstrates that the initiation rite is an active institution for girls of pubescent age. As alluded by a respondent of the case history in Embegueni Mzimba District:

> Our initiation rites have an emphasis on respecting men in our community, appreciating them and keeping our distance, and this is how we teach our girls to even sever their relationships with their fathers as they grow into women.
>
> (Personal communication of case history respondent in Mtelera village, Embegueni EPA)

In Tumbuka culture land allocation for crop enterprises is based on preference of cash crops that are perceived to be lucrative within the season; for example, soybeans are gaining prominence as a cash crop preferred by men and thus get larger allocations, as confirmed by respondents of the female FGD:

> Our husbands concentrate on crops that bring the most cash in a season, currently soybean pays more than groundnuts, thus they now allocate more land to soybean, groundnuts are still considered women's crops and we are given smaller plots.
>
> (Personal communication of FGD respondents in Chiwandauka-Chisi village, Emfeni EPA)

These qualitative findings indicate that women have an advantage relative over men in matrilineal-matrilocal households in terms of land rights as organized by kinship structures. However, control is through their brothers. Women in patrilineal-patrilocal households also do not have much agency, as more land is allocated to cash crops that earn more money than groundnuts.[4]

Gendered productivity differentials

This section presents pooled data estimates of groundnut productivity by gender and kinship structure using OLS regressions with bivariate and multivariate models. In both bivariate and multivariate models, the dependent variable was the groundnut yield per unit area (ha) (see Table 13.3). The first set presents the gender differences for Mchinji and Mzimba only, since there is no gender variation in Mangochi. However, the reference district is Mchinji. The second set presents differences based on all districts, with a comparison of women only and other independent covariates of productivity. Here the reference district is Mangochi.

The multivariate models in both sets present better coefficient results that are significant. The age of the farmers has a positive and significant effect on the yield according to the bivariate model of first set of models. Age is a proxy for experience in farming. In both models, higher yields from smaller plots reflect efficiency and ease of management. Kinship has a significant effect on groundnut productivity, as indicated by the coefficients in the multivariate model. As shown in Figure 13.2, farmers in patrilineal-patrilocal context (Mzimba) have significantly lower yields (570.78Kgs/ha) compared to farmers in matrilineal-matrilocal

Table 13.3 OLS estimates of groundnuts' productivity based on gender and kinship structure

Dependent Variable Yield/ha (Agricultural Productivity)				
	Gender differences (Mchinji, Mzimba only – Set 1		*District differences (women only) – Set 2*	
Variable	*Bivariate*	*Full model*	*Bivariate*	*Full model*
Age	4.963* (2.782)	8.052*** (3.305)	0.9588 (2.553)	3.414 (3.077)
Plot size (hectare)	−858.6** (207.9)	−831.50*** (212.4)	−836.05*** (208.5)	−1097.3*** (217.8)
District				
Mangochi/Matrilineal (yes = 1)	N/A	N/A	Reference	Reference
Mchinji/Matrilineal (yes = 1)	Reference	Reference	152.02 (78.17)	20.74 (115.8)
Mzimba/Patrilineal (yes = 1)	−196.6*** (77.93)	−332.4*** (86.92)	−162.2** (76.16)	−320.7*** (122.08)

Dependent Variable Yield/ha (Agricultural Productivity)				
	Gender differences (Mchinji, Mzimba only – Set 1		*District differences (women only) – Set 2*	
Variable	*Bivariate*	*Full model*	*Bivariate*	*Full model*
Literacy level (No. of years schooling)	−8.609	(14.21)	−0.424	(35.126)
	(11.34)	(12.48)	(10.37)	(16.849)
Soil fertility (fertile = 1; 0 otherwise)	275.7**	296.7***	276.1***	285.9***
	(84.38)	(83.12)	(81.63)	(80.63)
Constraints in accessing inputs (yes = 1)	−47.41	−75.46	-(65.72)	−63.88
	(80.28)	(74.95)	(76.67)	(72.58)
Extension service (yes = 1)	−45.88	48.16	−94.71	−23.97
	(80.40)	(83.26)	(76.08)	(82.62)
Received Training(yes = 1)	−220.9**	−135.3	−202.2**	−66.51
	(87.86)	(96.74)	(81.20)	(91.94)
Plot distance from homestead (m)	−0.072	−0.082**	−0.033	−0.043
	(0.0408)	(0.038)	(0.035)	(0.033)
Plot manager (male = base)				
plot manager (female)	−59.59	21.42	−32.34	0.035
	(80.84)	(86.21)	(75.38)	(95.50)
plot manager (joint)	154.23*	134.5	134.7	126.3
	(89.572)	(97.33)	(94.82)	(107.4)
Experience in farming g-nuts	2.258	2.258	7.676**	8.855**
	(3.898)	(3.898)	(3.609)	(3.99)
Constant		658.0		−320.7
		(229.2)		(214.4)
R-squared		0.215		0.186
Number of observations		212		284

Source: survey data Robust standard error in parenthesis *, **, *** Significant at 10%, 5% and 1%

context (Mchinji), as indicated by the coefficients in both sets of models. The higher the fertility of the plot, the higher the yields according to the coefficients in both sets of models. Plot distance from the homestead also significantly impacts on productivity as indicated by the coefficients in the multivariate model of the first set. The closer the plot to the homestead, the easier the access and hence management. Jointly managed plots have highest yields (Figure 13.3) in Mchinji. Experience in farming groundnuts also significantly affects productivity among women according to the bivariate and multivariate models in the second set.

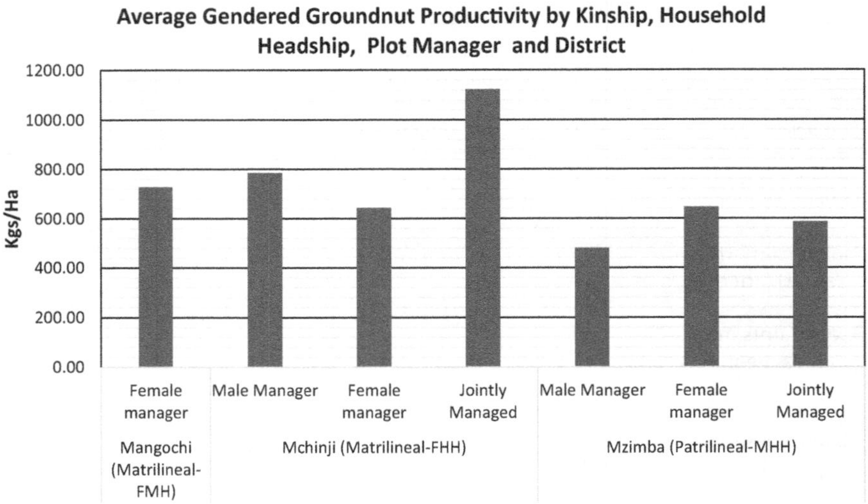

Figure 13.3 Average gendered groundnut productivity by kinship, household headship, plot manager and district

5. Discussion

We draw the following key observations from the findings. In matrilineal-matrilocal contexts, women have a disguised advantage over control of land, as their brothers have proxy authority overseeing sustainability of the lineage, households and agricultural enterprises. In patrilineal-patrilocal contexts, men continue to hold the primary authority and control over land, while women are relegated to subordinate positions of subsistence. In both contexts, male hegemony reinforces control over women, who also seem to have accepted this inequality. Systems of access to land and land rights, for both men and women, are embedded in kinship structures, which primarily organize social order, and consequently access to land is still largely through marriage. These findings resonate with the structuration theory by Giddens (1984) that human agency and social structure are in a continuous symbiotic relationship with each other. Actors use mutual knowledge based on instilled institutions to engage in practical action, and thus society is consciously reproduced and transformed in every social encounter, with resources and practices inherited from the past (duality of structure). It is thus critical for development interventions to understand social structure, cultural norms, beliefs and relational contexts to identify practicable and sustainable ways to improve livelihoods without exacerbating gender inequalities (cf. Kabeer, 2010; Cornwall, 2016). Hence, when designing interventions, a knowledgeable orientation on current gendered cultural reinforcements of context-specific layered rights to access and use of resources, roles and responsibilities is a prerequisite.

With regard to productivity, women in matrilineal-matrilocal contexts should be as productive as men. However, men in these contexts are producing more as the focus is entirely on groundnuts, a crop enterprise that hides their masculinity and averts meddling from the Nkhoswe. Groundnut enterprises seem to provide solace and space for men in matrilineal-matrilocal contexts and generate income as they work their way out of subordination. Thus, ownership and control of resources do not automatically translate to better productivity, as men in matrilineal-matrilocal contexts achieve higher productivity levels despite lack of land rights. Similarly, women in patrilineal households also produce more than men. Kinship structures place a cultural dividend on certain crops, and thus groundnuts are perceived as a "woman's crop" (cf. Orr et al., 2016),

Women in Mangochi have the highest productivity of all female managers despite lack of male labor, as their husbands have out-migrated. They solely manage their homes and sustain livelihoods through production of groundnuts among other crops. This finding unravels two important features – stability of marriages in matrilineal contexts and feminization of agriculture in Malawi. Besides out-migration, matrilineal marriages in Malawi are considered very unstable and have a high divorce rate (Jennifer, 2014). They are characterized as "walking marriages," as the man cannot claim any assets after a divorce or death of the spouse notwithstanding the bride service rendered (Kaarhus, 2010; Jennifer, 2014; Lovo, 2016). Feminization of agriculture is driven by out-migration of male labor, leading to women's increased responsibilities for household survival (cf. Lastarria-Cornhiel, 2006; Bikketi et al., 2016). Feminization of agriculture points to labor exploitation and worsening conditions for women, feminization of responsibilities taking on productive and reproductive roles (Lastarria-Cornhiel, 2006) and is framed as an invisible disempowering effect of labor in agriculture in FMH households (Bikketi et al., 2016).

In patrilineal-patrilocal contexts, priority of larger plots is given to other cash crops such as soybean, and thus women lack the agency to negotiate for larger plots for groundnut production due to hierarchical positions organized by kinship. Using Kabeer's critical moments framework, these findings point to institutionalized and invisible gender constraints on human action that give rise to durable forms of inequality that reproduce, modify or transform structural inequality (Kabeer, 2010). Men in matrilineal households and women in patrilineal households face gender-specific and gender-intensified constraints associated with norms and values of hegemony and exploitation, which disadvantage them in land ownership and other property rights (Kabeer, 2010). When actors are engaged in such patterns of practices, they draw resources of two kinds: authoritative ones (status) and allocative ones (resources). Giddens (1984) suggests that actors' capacity to carry out practices is influenced by their access to resources.

Lastly, the findings reveal an untapped youth potential in groundnut productivity in all regions based on the average age of farmers involved. According to OECD Development Center (2018), the youth population comprises 46% with the majority aged 15–29, accounting for one-quarter of the population in 2015, making Malawi a youthful country in the years to come.

6. Conclusions and policy implications

Key findings show that in all regions, cultural norms through kinship structures are still key institutions that determine gender norms and influence intra-household power relations in the three regions. They control residency and household headship, which in turn uniquely shape women's and men's agency, further determining productivity differentials and exacerbating inequalities in access and control of productive resources. As Giddens (1984 theorizes, social structure is the outcome of practices that have been routinized, making the same practices possible (dichotomy of structure), and is not separate from action. Thus it exists within memory traces of actors as a pattern of practices, and thus, as practices change, so does structure and vice-versa and is a product of repetition of the acts of individuals reproduced depending on the circumstances (Giddens, 1984). Hence, understanding of local realities, structures and opportunities that local contexts as routinized is a prerequisite for integrating ideologies, values and information to rationalize development interventions.

The gender system in both matrilineal and patrilineal contexts reveals the gender-specific and gender-intensified constraints (Kabeer, 2010) that the study posits are systemic components that are more resistant to change, impeding empowerment and agency of men and women in both matrilineal and patrilineal contexts, respectively. These systemic components impede agricultural development and must be addressed.

Agriculture is already Malawi's largest employer and is uniquely positioned to absorb the entry of large numbers of youth to accelerate productivity given the global demand for food security (Brooks et al., 2013). Thus, it is crucial to develop policies and strategies to harness the untapped youth dividend in agricultural development.

Feminization of agriculture in the southern region is on the rise with the high out-migration of men as an escape from subordination by matrilineal customs. This gendered agrarian change can lead to two scenarios: (1) the potential of increasing deficits in overall agricultural productivity due to less participation in agricultural production because of lack of male labor and increased workload and (2) abandoning agriculture due to reliance on remittances. Either way, it leads to underperformance in agricultural productivity. Women are taking on greater responsibilities for agricultural production, in addition to reproductive responsibilities, further exacerbating obstacles such as time constraints and health hazards, among others, that lead to less productivity. Overall, there are implications for development policy and strategies when rationalizing interventions in Malawi as kinship structures are routinized. These are (1) the extent of agricultural development investments should explicitly target women, men or youth to redress gender imbalances and (2) whether a woman-focused or a youth-focused agricultural investment strategy could yield a double dividend – increasing gender equity at the same time as increasing overall productivity (Doss, 2017).

Acknowledgments

The authors are grateful to the CGIAR Gender Network and International Crops Research Institute of the Semi-Aris Tropics (ICRISAT) for funding this study under the CGIAR Collaborative Platform for Gender Research.

References

Bikketi, E., Chinwe, I.S., Bieri, S., Haller, T. and Wiesmann U. (2016). Gendered division of labour and feminisation of responsibilities in Kenya; Implications for development interventions. *Gender, Place & Culture*, 23(10), pp. 1432–1449. doi:10.1080/0966369X.2016.1204996.

Brooks, K., Zoray, S., Gautam, A. and Goyal A. (2013). Agriculture as a sector of opportunity for young people in Africa. *World Bank Policy Research Working Paper* no. 6473, pp. 1–43. doi:10.1596/1813-9450-6473.

Cornwall, A. (2016). Women's empowerment: What works. *Journal of International Development*, 96(28), pp. 342–349. doi:10.1002/jid.

Dean, B. (2015). Brideservice. In international encyclopedia of anthropology. *Research Gate*, pp. 1–4.

Doss, C. (2017). Women and agricultural productivity: Reframing the Issues. *Development Policy Review*, pp. 1–16. doi:10.1111/dpr.12243.

Doss, C. (2013). *Intrahousehold bargaining and resource allocation in developing countries*. Working Paper, Oxford University Press for World Bank Research Observer. doi:10.1093/wbro/lkt001.

Food and Agriculture Organisation of the United Nations (FAO). (2015). *Country fact sheet on food and agriculture policy trends in Malawi*. Available at: http://www.fao.org/3/a-i4491e.pdf.

Giddens, A. (1984). The constitution of society: Outline of the theory of structuration. *Cognitive Therapy and Research*, 12(4), p. 448. doi:10.1007/BF01173303.

Jennifer, K. (2014). *Shifting roles in gender, kinship and household: Women's empowerment in Matrilineal Malawi*. Doctoral Thesis, Emory University.

Kaarhus, R. (2010). Women's land rights and land tenure reforms in Malawi: What difference does matriliny make? *Forum for Development Studies*, 37(2), pp. 171–192. doi:10.1080/08039411003725857.

Kabeer, N. (2010). Women's empowerment, development interventions and the management of information flows. *Ids Bulletin-Institute of Development Studies*, 41(6), pp. 105–113. doi:10.1111/j.1759-5436.2010.00188.x.

Lastarria-Cornhiel, S. (2006). *Feminization of agriculture: Trends and driving forces*. Background Paper for the World Development Report 2008, 1–26. Wisconsin, Madison: Rimisp-latin American Centre for Rural Development.

Lovo, S. (2016). Tenure insecurity and investment in soil conservation. Evidence from Malawi. *World Development*, 78, pp. 219–229. doi:10.1016/j.worlddev.2015.10.023.

Matinga, B. (2015). *What's mine? Rural women's experiences around property rights in the context of dissolved marriages in Matrilineal societies: A case study of Muluwila Village in Kuntumanje Area, Zomba District, Malawi*. Doctoral Thesis, Stellenbosch University.

Meijer, S.S., Catacutan, D., Ajayi, D.C., Seleshi, W.D. and Nieuwenhuis, M. (2015). The role of gender and kinship structure in household decision-making for agriculture and tree planting in Malawi. *Journal of Gender, Agriculture and Food Security*, 1(1), pp. 54–76.

Mushibwe, C. (2009). *What are the effects of cultural traditions on the education of women? (The study of the Tumbuka people of Zambia)*. Doctoral Thesis, University of Hudderesfield.

Mwambene, L. (2005). *Divorce in Matrilineal customary law marriage in Malawi: A comparative analysis with the patrilineal customary law marriage in South*. Law Thesis, University of West Cape.

OECD Development Center. (2018). *Youth well-being policy review of Malawi*. EU-OECD Youth Inclusion Project Paris, p. 148.

Orr, A., Takuji, T., Kee-Tui, H.S. and Msere, H. (2016). What do we mean by women's crops'? Commercialisation, gender and the power to name. *Journal of International Development*, 53(9), pp. 1689–1699. doi:10.1017/CBO9781107415324.004.

Peters, P. (2012). Against the odds; Matriliny, land and gender in the Shire highlands of Malawi. *Harvard University*, 28(23), pp. 1–21.

Peters, P.E. (2010). "Our daughters inherit our land, but our sons use their wives' fields": Matrilineal-matrilocal land tenure and the new land policy in Malawi. *Journal of Eastern African Studies*, 4(1), pp. 179–199. doi:10.1080/17531050903556717.

Rao, S. (2016). *Indicators of gendered control over agricultural resources*. A guide for Agricultural Research and Policy Working Paper, CGIAR Gender and Agriculture Research Network. CGIAR Consortium Office and International Center for Tropical Agriculture (CIAT). Cali, Colombia, p. 74.

Slavchevska, V., De la O Campos, P.A., Brunelli, C. and Doss, C. (2017). Beyond ownership: Women's and men's land rights in Sub- Saharan Africa. *Responsible Land Governance: Towards an Evidence Based Approach*, pp. 1–32. Available at: www.conftool.com/landandpoverty2017/.

Takane, T. (2008). Customary land tenure, inheritance rules, and smallholder farmers in Malawi. *Journal of Southern African Studies*, 34(2), pp. 269–291. doi:10.1080/03057070802037969.

Takuji, T., Msere, H., Siambi, M., Mazvimavi, K. and Okori, P. (2016). Evolution and impacts of groundnut research and development in Malawi: An ex-post analysis. *African Journal of Agricultural Research*, 11(3), pp. 139–158. doi:10.5897/AJAR2015.10167.

Terrell, S. (2012). Mixed methods research methodologies. *The Qualitative Report*, 17(1), pp. 254–280.

World Development Report. (2012). *Gender equality and development*. Washington, DC: The International Bank for Reconstruction and Development, The World Bank, pp. 72–91.

14 Changes in participation of women in rice value chains

Implications for control over decision making

Sujata Ganguly, Leif Jensen, Samarendu Mohanty, Sugandha Munshi, Arindam Samaddar, Swati Nayak and Prakashan Chellattan Veettil

Introduction

Women have always played an integral, although often unrecognized, role in agriculture in India and around the world. As has been underscored by the extensive and growing body of literature on gender and agriculture (FAO, 2011), women's roles and responsibilities in agriculture – within the household and the community – are complex, diverse and multifaceted (Doss, 2001; Doss and Morris, 2001; Quisumbing and Pandolfelli, 2010; Meinzen-Dick et al., 2011). Women play a significant and crucial role in agricultural development and allied fields like the main crop production, livestock production, horticulture, post-harvest operations, agro/social forestry, fisheries etc. The nature and extent of women's involvement in agriculture, no doubt, varies greatly from region to region. Studies suggest that, despite the wide variability among regions and countries, women farmers play a major role in labor-intensive activities such as planting and weeding, among many other agricultural activities (Meinzen-Dick et al., 2011). With the exception perhaps of plowing, women are actively involved in almost all agricultural production work.

The feminization of agricultural work results in part from the coping strategies adopted by rural families to increase and diversify incomes and livelihoods, and usually as an economy matures, there is a movement of excess agricultural workers from low-productivity agriculture to higher-productivity sectors like manufacturing and services and thus from rural to urban areas and from lower wages to higher wages. Further, the coping strategies are gendered: men usually work with lucrative crops or migrate as seasonal or permanent workers; while women cultivate the family plot for household consumption, care for small livestock and process and/or sell part of their production in local markets (Kim and Bishop-Sambrook, 2014). When women are left behind due to male out-migration, it is assumed that they get control over land and even over choices relating to production (Feldman, 2009). Male out-migration too has consequences on the livelihood of rice farming households in the place of origin and on the women left behind. Remittances may be used to pay for farm inputs, including wage laborers to compensate for the absence of male family members. If remittances are not large enough, women's workload may increase. In the absence of men, women's decision-making authority may increase, and their roles may shift from unpaid family helper to decision makers.

As per FICCI (2015), the size of the workforce in Indian agriculture declined by around 30.57 million between 2004–05 and 2011–12, although the size of the total workforce increased. There has been less diversification in the occupations of rural women and their movement away from agriculture when compared to their male counterparts. Hence, this shift in male occupation has implications for women as their increased participation in agricultural activities may or may not increase the intra-household decision-making power. Also, in the process of "empowering" women are women overburdened? Further, women's participation in decision making on land use and income, expenditure depends on intra-household dynamics. Apart from Southeast Asia, Asian women have little decision-making power at the household level, including on the use of their own earnings (Kim and Bishop-Sambrook, 2014). Kandiyoti (1988) outlines contexts of "classic patriarchy" – patrilocal extended households where a senior man is in control of the family, including younger men.

The eastern Indo-Gangetic Plains in India are dominated by small farm sizes, low incomes and comparatively low levels of agricultural mechanization, irrigation and productivity. Traditionally, farm women have tended to be invisible workers, with their contributions to their farms, families, communities and the larger economy as a whole unacknowledged. It leads to the triple burden of reproductive, productive and community management (the social spheres) work for women (Grassi et al., 2015). Because they toil in the background and because their work involves multiple overlapping roles that are not easily delineated, the lived experience of rural and farm women has not been documented to any great extent. Hence, an attempt has been made to understand the changes in gender participation in rice value chain firstly with emphasis on gender roles in specific agricultural activities (land preparation, crop establishment, sowing, transplanting), crop care (fertilizer, pest management, weeding), harvest, post-harvest (threshing, drying, storage and marketing) and secondly to understand the intra-household decision making in agriculture.

Research methodology

The Indo-Gangetic Plain is one of the world's major food grain–producing regions. During the past 30 years, agricultural production growth in this region has been able to keep pace with population demand for food in the country mainly due to the adoption of green-revolution technologies inducing yield growth, followed by area expansion. But this opportunity is ceasing very fast due to limited scope for increasing the availability of arable land and natural resources (Singh et al., 2011). This research was conducted in 2015 in three states of eastern India – Bihar, Odisha and West Bengal. Within each state, districts in which more than 50% of all the agricultural output is in rice production were first identified. Among these, five districts per state (for a total of 15) were selected purposively to represent different ecological zones and irrigation statuses. In each district, the top two rice-growing blocks[1] were chosen with a total of 30 blocks. In each block, five villages were randomly selected, for a total of 150 villages per state. Finally, within sampled villages, a census was conducted to identify rice-farming households including their caste status. From this list, stratified random sampling was used to ensure proportionate representation of caste groups. The total sample size is 1,484 women farmers.

The woman rice farmer in each household was asked about their current participation and extent of decision-making authority related to different agricultural activities and what it was 10 years ago. There were 14 statements related to gender norms and perceptions that were asked based on five-point Likert scale (ranging from 1 = strongly disagree to 5 = strongly agree). Using factor analysis, the 14 statements formed three broad categories namely, access, identity and recognition; access to and awareness of economic resources; and community and household perceptions. These norms have been asked with reference to two time periods, i.e., the current scenario and the situation that prevailed 10 years ago.

Results

The sampled households are mostly Hindu (93%). One-third belong to the general caste category (33%), another third belong to other backward classes[2] (35%), while 24% belong to the scheduled caste category and 9% to the scheduled tribe. The majority of the respondents are the wife of the household heads, while 7% respondents are the heads of the households. The age distribution suggests a relatively young population with over half of the respondents (56%) under 40 years old. A majority (58%) are illiterate; they do not have formal education or have a primary level of education. On average, the respondents have 13 years of experience in agriculture. The households with out-migrant members are 20%.

The changes in the last decade

The changes in the last decade are assessed based on three areas i.e., sources of household income, social network and access to agriculture extension and information services. It is believed that restrictive gender norms that constrain women's interactions with men and their access to formal education and training, they necessarily are blocked from information, institutions and markets (Primo, 2003). However, in a study conducted in 2016 by Ganguly et al. in Bihar and Odisha, it was found that women are getting exposed to trainings related to health, agriculture etc. and their affiliations to self-help groups have enabled them to move out of their orbit of confinement. Kandiyoti (1988) outlines contexts of "classic patriarchy" – patrilocal extended households where a senior man is in control of the family, including younger men. Women expect, over time, to be able to exercise power based on their status as mothers of sons and mothers-in-law, a situation whereby older women have gains to make from the subordination of younger women. Kandiyoti argues that this system leads women to maximize their own life chances through collusion in, and reproduction of, their own subordination – the "patriarchal bargain.'"

The changes in sources of household income in the last decade: The households reporting earnings from their own businesses (self-employed) have increased from 9% to 13% as primary source of household income and 8% to 12% as secondary source of household income over the period of 10 years. The earnings of the household members who are working as agricultural laborers which they consider as their primary source of household income, has decreased from 8% to 6% over the period of 10 years, and those that consider as their secondary source of household income, has decreased from 13% to 11% over the period of 10 years. The households reporting remittances as primary source of income have increased from around 2% to 6% over the period of 10 years. This may be due to the fact that the rate of migration from the states to both rural and urban destinations is very high and is believed to have increased over the decade, with rural people becoming more mobile (Deshingkar and Akter, 2009). Moreover, remittances from migration have significantly contributed to increased incomes in the states (Bhaskaran and Mehta, 2009). Earlier 24% of the households had no secondary source of income while currently it is 12%. This reveals that over the period of 10 years, people have alternate sources of income. Also, salaried employment and cultivation (own) as secondary sources of household income have increased over the period of time. Interestingly, cultivation as a primary source of income decreased in the last decade (Table 14.1). This reflects that people have gradually started opting for non-farm activities as a part of their livelihood strategies.

Table 14.1 Changes in the sources of primary and secondary household income

Sources of household income (N = 1484)	*Primary source of income (%)*			*Secondary source of income (%)*		
	Past 10 years	*Current year (2015)*	$t^{\#}$	*Past 10 years*	*Current year (2015)*	$t^{\#}$
None	0.4	0.1	−1.9	23.8	12.1	-11.5^{***}
Self-employed	9.4	12.5	4.0^{***}	8.0	11.7	4.4^{***}
Livestock/ Poultry Rearing/ Fishery	0.9	1.2	0.8	10.2	9.2	−1.4
Salaried employment	4.0	4.2	0.4	1.2	2.0	1.8
Agriculture/ Cultivator/ Farmer	60.2	53.8	-5.4^{***}	28.6	37.5	7.7^{***}
Agricultural labor	8.2	5.9	-3.7^{***}	13.1	10.5	-3.6^{***}
Non-Agricultural labor	13.5	14.6	1.4	11.5	11.0	−0.6
Remittance	2.4	6.1	6.8^{***}	3.1	5.3	3.3^{***}
Pension	0.3	0.9	2.7^{***}	0.3	0.5	1.6

Note: $^{\#}$paired *t*-test; Significance level $^{***}p < 0.01$.

The changes in social network in the last decade: Over the period of 10 years, household members' social network or institutional affiliation has increased from 17% to 36%. There is statistically significant increase in the women's affiliation to the organizations while there is decrease in the affiliation to organizations among men in the households but this decline is not statistically significant (Table 14.2). The women are mostly associated with self-help groups while the men are associated mostly with farmers' clubs and cooperatives.

The changes in access to agriculture extension and information services in the last decade: The household members' access to agriculture extension and information services has significantly increased over the period of 10 years i.e., 35% of households have received agriculture extension and information services 10 years ago while 53% of households have received the same in the past 12 months. The women's access to information has increased from 7% to 17%, while that of men has increased from 33% to 50% over the period of 10 years (Table 14.3). The important sources of information are input dealers and agriculture extension agents. This increase in the access to agriculture extension and information services can be attributed to the increased number of projects and information sources over the period of time. Access to information by both men and women is very key to any development and welfare of the society. A study on climate-smart agricultural practices showed that, even though women tend to be less aware of these practices than men, they are no less likely to adopt them once they gain access to new knowledge (Jost et al., 2015).

Table 14.2 Changes in social network

Changes in the last decade (N = 1484)	*Past 10 years (%)*	*Current year (2015) (%)*	$t^{\#}$
Household members' affiliations to organizations	17.3	36.3	15.1^{***}
Women's affiliations to organizations	13.4	32.8	15.7^{***}
Men's affiliations to organizations	26.4	21.4	−0.9

Note: #paired *t*-test; Significance level $^{***}p < 0.01$.

Table 14.3 Changes in access to agriculture extension and information services

Changes in the last decade (N = 1484)	*Past 10 years (%)*	*Current year (2015) (%)*	$t^{\#}$
Household members' access to agriculture extension and information services	35.2	53.3	15.4^{***}
Women's access to agriculture extension and information services	7.4	16.5	10.3^{***}
Men's access to agriculture-related information	32.8	49.6	13.8^{***}

Note: #paired *t*-test; Significance level $^{***}p < 0.01$.

Changing gender norms and perceptions

Gender norms are powerful, pervasive values and attitudes about gender-based social roles and behaviors that are deeply embedded in social structures. Gender norms manifest at various levels, including within households and families, communities, neighborhoods and wider society. Understanding how this framework operates at the household level is important because it provides a sort of mental map of the acceptable roles, responsibilities and behaviors for each household member. Figure 14.1 shows that there has been a change in perception over the past 10 years. The women feel that as compared to the past, they cannot directly deny that women have equal access to training in agriculture, have knowledge of agricultural technologies, are involved in mechanized agricultural operations, identified/recognized as farmers or regarding the women wage rate compared to men. Ten years back, the situation was not in favor of women, but currently, the respondents have given neutral responses. This depicts that the respondents are neither denying that at present they do not have access, identity and recognition, nor are they accepting it. From the section above, there has been an increase in access to information related to agriculture as well as affiliation to organizations, but there is further need of improvement, and programs need to target this aspect of women's status.

When it comes to access to economic resources, i.e., whether women are informed about debts or monetary transactions of the family, 10 years back, the situation was indecisive while currently the situation seems to have improved, and women agree that they do have access to economic resources. There has been improvement in community and household perception. Ten years back, the situation was indecisive, while now the women agree that community and household perceptions are positive. Overall, there is a positive shift in perceptions as the differences are statistically significant (by using paired *t*-test).

Analysis of the 14 perception statements separately reveals that women report a positive shift with regard to household decision making and that they are informed about debts or any monetary transactions in the family. Regarding women's equal access to training in agriculture, involvement in mechanized

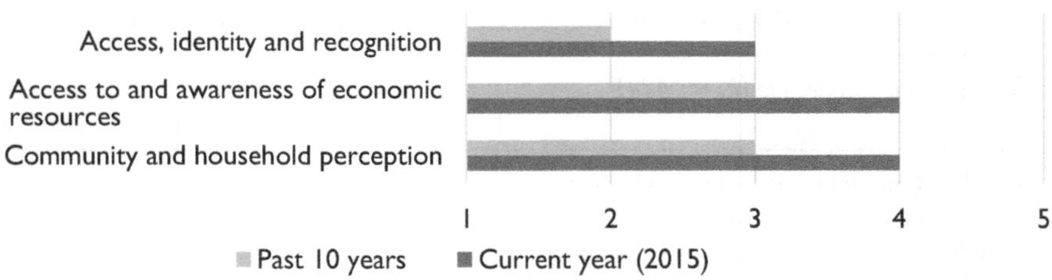

Figure 14.1 Mean change in norms and perceptions in the last decade and current scenario

Note: The mean value depicts the 5 point Likert scale used to understand the norms, 1-Stongly disagree to 5-Strongly agree.

agricultural operations and recognition as "farmers," they are indecisive at present. However, 10 years back they disagree that they had such trainings, involvements or recognitions. Women are indecisive regarding whether women have knowledge about agricultural technologies at both the time periods. However, they do agree that they are better managers of farm if they get equal opportunity and access to resources while they were neutral when referred 10 years back.

The women confirm that farm mechanization reduces drudgery. The women agree that they work off-farm to ensure their families' food security while when referred 10 years back regarding the same, they are quite uncertain. There is a disagreement regarding wage rate and land ownership. They agree that there is discrimination in wage rate as well as land ownership as compared to men. Land ownership and access is quite a debatable gender issue. As stated by Sultana (2011), the struggle over access and control of resources are products of individual needs and decisions as well as a multitude of other factors such as institutions, relationships and emotions.

Gendered participation in agricultural activities

Agricultural workers have constituted almost 75% of the total rural working population as of 2011, from which 45% and 55% are listed as household cultivators and hired agricultural laborers, respectively (Ministry of Agriculture, 2015). Hired agricultural laborers are to be distinguished from household cultivators, where hired agricultural laborers denote those rural workers who are employed on wages and household cultivators are the employers of hired agricultural laborers. However, seasonally, as small owners and tenants find work as wage-dependent agricultural laborers, the line between cultivators and agricultural laborers is subject to marginal shifts, wherein the agricultural workers may be broadly described "as a residuary group in the rural community" (Planning Commission, 1952). Based on the aforementioned definitions, women farmers account for almost 30% of the cultivators and 40% of the agricultural laborers as of 2011 (Ministry of Agriculture, 2015). In many countries, the role of women in agriculture is considered just to be a "*help*" and not an important economic contribution to agricultural production (Hill, 2011; Mucavele, 2013). This implies that women are still underrepresented in agricultural practices, although participation greatly varies across different crops. Participation means that people are closely involved in the economic, social, cultural and political processes that affect their lives. Since participation requires increased influence and control, it also demands increased empowerment in economic, social and political terms (UNDP, 1993).

Table 14.4 indicates that men's participation in the agricultural activities has shown a positive trend, i.e., men continue to participate in different agricultural activities. The percentage change in land preparation, crop establishment, post-harvest and marketing from 10 years back to present is around 3 while around 2 in crop care and 4 in harvesting. Though the percentage change is positive, the magnitude of change is small. One reason could be that as men have been traditionally involved in agriculture, their roles and responsibilities did not see a

Table 14.4 Percentage point change in participation in agricultural activities in the last decade

Activities	*Percent point change from past to present*			
	Male members	*Female members*	*Hired male laborers*	*Hired female laborers*
Land preparation/Seedbed preparation	3.2***	−6.6	4.5**	21.5
Crop establishment	2.8***	−7.4***	9.6***	12.9***
Crop care	2.4***	−2.3	9.8***	16.4***
Harvesting	3.6***	−1.6	9.1***	14.1***
Post-harvest	2.9***	4.5	9.6***	14.2***
Marketing	3.2***	21.3**	11.4**	12.5

Note: Significance level ***$p < 0.01$ **$p < 0.05$ by using paired t-test.

vast change. They were involved in agriculture in the past as well as the present. However, in the era where we talk about "feminization of agriculture," we see that there is decrease in the participation of the women in different agricultural activities except post-harvest and marketing. One possibility can be that now women have entered into a role of hiring and supervision of the hired laborers rather than getting involved in different activities or that the increasing farm mechanization have led to decreasing women's participation in agriculture.

In marketing, there has been a positive change of 21% in female participation, while in post-harvest activities, the change is around 5%. Over the period of 10 years, the percentage change in hiring male or female laborers is also positive. This reveals that hiring of laborers has increased. The hiring of female laborers is higher than male laborers. This confirms to a certain extent the finding stating that there has been decreased participation of women in agriculture.

In Table 14.1, it was observed that "agricultural laborers" as source of income has decreased, but Table 14.4 depicts that the percentage change in hiring laborers, irrespective of gender, has increased. The plausible reason can be that as majorly female laborers have shown change, their wages may not be considerable enough to be a source of income neither as primary nor as secondary, rather remittance or income from non-farm activities can be assumed to gain greater importance.

Gendered decision-making authority in agricultural activities

The male dominance in decision making in the household and economy has continued even in areas where women are the key providers of labor because the influence of women has not been recognized. To reword Doss (2001) but in another regard, the majority of women farmers live in male-headed households and farm on male-owned plots. Access to resources does not guarantee control over them (cited in Ngenzebuke, 2014), since the one in control might dominate in the decision making. From the aforementioned sections, it is revealed that women have decreasing participation in land preparation, crop establishment, crop care and harvesting but an increasing participation in post-harvest activities and marketing. It is imperative to understand their decision-making authority in these activities. Though there has been a positive shift in marketing, do the women have decision-making authority regarding the same?

The respondents were asked about their involvement in decision making related to different agricultural activities and whether men and women in the households were involved while taking these decisions. Table 14.5, which depicts that percentage change from past 10 years to present, shows a positive trend for the respondent themselves and for the men. However, the magnitude of change is lower among men than women. The percentage change in decision making from past to present in different agricultural activities has been between 45% and 49% for women, while for men, the change has been around 3% for different agricultural activities. This can be due to the fact that traditionally men had been taking decisions and hence the percentage is not drastic, while the women have gradually entered this sphere of "agency" and hence a positive and considerable change. For other female members, the trend is negative. Here, the other women relate to mothers-in-law, daughters-in-law, sisters-in-law, adult daughters or any other adult women.

Assessing the factors affecting the changes in the women's decision-making power

Intersectional analysis aims to reveal multiple identities, exposing the different types of discrimination and disadvantage that occur as a consequence of the combination of identities. While it is assumed that women face discrimination in one sphere or the other, factors like caste, religion, household composition, sources of income, male out-migration from the households, whether a woman is the head of the household, age, marital status, level of educational attainment, institutional or organizational affiliation, access to agriculture extension and

Table 14.5 Percentage point change in decision making in the last decade

	Percent point change from past to present		
Decision-making authority	*Self*	*Male members*	*Other female members*
Land preparation/Seedbed preparation	44.8***	3.5***	−26.8***
Crop establishment	47.7***	3.4***	−24.2***
Crop care	46.4***	2.7***	−13.1***
Harvest/Post-harvest	46.7***	2.8***	−17.3***
Marketing	49.2***	2.2**	−25.8***

Note: Significance level ***$p < 0.01$ **$p < 0.05$.

information services, women's participation in different agricultural activities, men's participation in different agricultural activities and perception regarding different community norms have been considered in this section.

Women and men engage in different farm activities. Women generally lack access to the factor of production, and this is informed by gender and cultural norms, gendered institutions and practices which privileges men. While not all men enjoy the same gender privileges, not all women are subordinated, in the sense this research moves away from the perspective which view all women are disadvantaged. Table 14.6 reveals that the women belonging to other backward classes are 1.6 times more likely to have positive change in decision making as compared to scheduled caste. With increasing women's participation in different

Table 14.6 Factors affecting changes in women's decision making over the period of 10 years

	Exp(B)	*Sig.*
Household variables		
Caste		0.072
Scheduled caste®		
Scheduled tribe	0.56	0.176
Other backward classes	1.55	0.077
General	1.10	0.767
Changes in 10 years		
Change in women's participation in agricultural activities	1.22	0.004
Change in men's participation in agricultural activities	0.83	0.013
Perceptions		
Women are involved in household decision-making process	1.06	0.574
Women have equal access to training in agriculture	1.03	0.675
Women have knowledge of agricultural technologies	0.91	0.258
Women are informed about debts or monetary transactions of the family	0.78	0.004
Women are better managers of farm if they get equal opportunity and access to resources	0.98	0.761
Women are involved in mechanized agricultural operations	0.96	0.618
Rural women work off-farm to ensure their families' food security	1.27	0.012
Farm mechanization reduces drudgery of women	1.03	0.733
Women are identified/recognized as farmers	1.07	0.356
For same work, women wage rate is higher than men wage rate	1.14	0.077
Women having access to economic resources prioritize household welfare investments	0.99	0.916

Note: Dependent variable: Women reporting <U> no change in decision making in agricultural activities in the last decade = 0; Women reporting positive change in decision making in agricultural activities in the last decade = 1;

Independent variables: The other variables considered as factors and are not significant are religion; total men and women in the household aged 15–60 years; primary source of income; out-migrants in the household; whether respondent is household head; age of the respondent; marital status; years of experience in agriculture; level of educational attainment; changes in institutional or organizational affiliation, access to agriculture extension and information services.

agricultural activities, women are 1.2 times more likely to have positive change in decision making, while with increasing men's participation in different agricultural activities, women are 17% less likely to have positive change in decision making.

Those who perceive that women are informed about debts or monetary transactions of the family are 22% less likely to have experienced positive change in decision making, while those who perceive that rural women work off-farm to ensure their families' food security are 1.3 times more likely to have positive change in decision making, while those who believe for same work, women's wage rate is higher than men's wage rate, are 1.1 times more likely to have experienced positive change in decision making.

Another set of binary logistic model for positive change in participation (did not participate in the past but currently participating) depicts that women are 3.5 times more likely to report positive change in participation in agricultural activities with increasing number of out-migrants in the households. Also, there is an inverse relation between decision making and participation in agricultural activities, i.e., the higher the women reporting positive change in decision making in agricultural activities, the lower is the participation (table not shown). Does this mean that when women get involved in decision making, they gradually cease to participate in agriculture rather they start managing and supervising the hired laborers?

Discussion

Understanding the context, connections and circumstances is important to infer the way people relate to one another and negotiate their responsibilities and relationships. In order to make some meaningful interpretation of the changing role of women in rice production, changes in some of the background characteristics were assessed. For example, it can be seen that agricultural laborers are decreasing while remittance and self-employment as sources of income are increasing. Can this be interpreted as a positive outcome or a negative outcome? I mean positive in the sense that people have avenues to switch their occupation from agricultural laborers to some other occupations or negative in the sense that farm laborers are decreasing, which can have inadvertent impact on farming. Usually, as an economy matures, there is a movement of excess agricultural workers from low-productivity agriculture to higher-productivity sectors like manufacturing and services, and thus from rural to urban areas and from lower wages to higher wages. Also, whether people prefer to out-migrate in search of jobs or start their own businesses is an emerging question and scope for further research. Also, higher numbers of people have alternate sources of income as compared to the past.

The women have greater access to information related to agriculture compared to the past. The important sources of information are input dealers and agriculture extension agents. Agricultural extension strategies traditionally have targeted on men for training, information and access to inputs and services. Second, women's daily workloads do not usually allow them to be absent from home. Indeed, even attending short courses may cause insuperable problems in arranging substitute care for children or the home. Further, in the overwhelming majority of countries, extension services have been staffed predominantly by men. The women have greater institutional affiliation than what it was in the past. The women are mostly associated with self-help groups, while the men are associated mostly with farmers' clubs and cooperatives. Empowering women starts with building their self-confidence. Experiences across the region have shown that self-help groups (SHGs) are key to building women's confidence, voice and bargaining power.

Apart from the economic and social opportunities, it is pertinent that we discuss the existing norms and perceptions and the changes in the last decade to be able to infer about the changing role of women comprehensively. It was observed that women have started getting access to economic resources. There has been a positive shift with regard to household decision making and that women in the households are informed about debts or any monetary transactions. Women agree that they are better managers of farms if they get equal opportunity and access to resources and confirm that farm mechanization reduces drudgery. The women agree that they work off-farm to ensure their families' food security, while when referred 10 years back regarding the same, they are quite uncertain. There is disagreement regarding wage rate and land ownership. They agree that there is discrimination in wage rate as compared to men as well as land ownership.

In the present study, the men continue to participate in different agricultural activities. In the era where we talk about "feminization of agriculture," we see that there is decrease in the participation of the women in different agricultural activities (Sanghi et al., 2015) except post-harvest and marketing. There seems to be an emerging opportunity for women to access the market. However, the scope for further research remains whether their access to market is limited to physically going and selling the products without having any control of the income. It is of importance to all rural households, in improving their access to markets and must be considered a critical element of any strategy to enable them to enhance their food security and increase their incomes. However, the question remains: does the decreased participation in agricultural activities lead to further decrease in decision-making authority? The study shows that there is a positive change in decision making for both men and women, but the magnitude of change is higher for women than men. As pointed out by Kabeer (1999), we can also distinguish between various critical "control points" within the decision-making process itself where such control is defined in terms of the consequential significance of influencing outcomes at these different points. This can be cited as one of the limitations of the chapter that these "control points" cannot be identified to get a holistic picture of decision-making authority of women in the rice value chain.

The constraints that are common to all family farmers are often exacerbated for rural women because restrictive customary norms on women's role in households and public life limit their ability to make decisions and seize opportunities (Kim and Bishop-Sambrook, 2014). Accordingly, the main predictors of change in the last decade can be cited as the women in other backward classes have positive change in decision making than the women in scheduled castes. Increased men's participation decreases decision-making authority of women, while there is a direct relation with women's participation and decision making. The women who work off-farm and those who are positive about wage rate difference between men and women have reported positive change in decision making in the last decade.

Conclusion

The study concludes that even though we see decreased women's participation in agricultural activities except post-harvest and marketing and increased decision-making authority, does it truly reflect that women are now having emerging opportunities to "decide?" There is further scope to delve deep into this aspect of "decision making" and understand the extent to which they are involved and heard while taking any decision as women face many constraints in the multiple activities they pursue – less land ownership, access to loans, extension and other services and ability to hire laborers (Doss et al., 2008). The increased participation in marketing raises a question of whether the women have entered the "male domain" and have got access to the markets. If it is so, then what are the triggers and barriers that the women faced? The increased participation of hired male and female laborers requires further in-depth understanding of the fact of who decides and who manages the hired laborers in the households. The deeper we delve in the subject, farther seems the horizon where we will get the answer to all our queries. However, the study is just the beginning that raises some of the pertinent questions to understand the changing role of women in rice production over the period of time.

Notes

1 The community development block is a rural area earmarked for administration and development in India. The area is administered by a Block Development Officer. A community development block covers several *gram panchayats*, local administrative units at the village level.

2 The Indian Constitution declared certain castes and tribes Scheduled Castes (SCs) and Scheduled Tribes (STs) to create a system of affirmative action in employment and education for these groups. Other Backward Classes (OBCs) are a group of castes that are also supposed to be backward in respect of socio-economic development, but their social position is better than STs and SCs.

References

Bhaskaran, R. and Mehta, B. (2009). *Tracing migrants in Delhi from Bihar-An enquiry on the role of migration as a development facilitator in poor origin areas*. New Delhi: Institute for Human Development and SANEI.

Deshingkar, P. and Akter, A. (2009). *Migration and human development in India*. Human Development Research Paper 2009/13. New York: United Nations Development Programme.

Doss, C. (2001). Designing agricultural technology for African women farmers: Lessons from 25 years of experience. *World Development*, 29(12), pp. 2075–2092.

Doss, C., Grown, C. and Deere, C. (2008). *Gender and asset ownership: A guide to collecting individual-level data*. Policy Research Working Paper 4704. World Bank.

Doss, C. and Morris, M. (2001). How does gender affect the adoption of agricultural innovations? The case of improved maize technology in Ghana. *Agricultural Economics*, 25, pp. 27–39.

FAO. (2011). *The state of food and agriculture 2010–2011. Women in agriculture: Closing the gender gap for development*. Rome, Italy: Food and Agriculture Organization of the United Nations.

Federation of Indian Chambers of Commerce & Industry (FICCI). (2015). *Labour in Indian agriculture: A growing challenge*. Available at: http://ficci.in/spdocument/20550/FICCI-agri-Report%2009-03-2015.pdf [Accessed 1 December 2016].

Feldman, S. (2009). Historicizing garment manufacturing in Bangladesh: Gender, generation, and new regulatory regime. *Journal of International Women's Studies*, 11(1), pp. 268–288.

Ganguly, S., Munshi, S. and Nayak, S. (2016). *Women on the path to progress*. Available at: http://ricetoday.irri.org/women-on-the-path-of-progress/ [Accessed 10 September 2016].

Grassi, F., Landberg, J. and Huyer, S. (2015). *Running out of time: The reduction of women's work burden in agricultural production*. Rome, Italy: Food and Agriculture Organization of the United Nations.

Hill, C. (2011). *Enabling rural women's economic empowerment: Institutions, opportunities, and participation*. UN Women In cooperation with FAO, IFAD and WFP Expert Group Meeting 'Enabling Rural Women's Economic Empowerment: Institutions, Opportunities and Participation.' Available at: www.un.org/womenwatch/daw/csw/csw56/egm/Hill-BP-1-EGM-RW-Sep-2011_CH.pdf [Accessed 15 June 2016].

Jost, C., Kyazze, F., Naab, J., Neelormi, S., Kinyangi, J., Zougmore, R., Aggarwal, P., Bhatta, G., Chaudhury, M., Tapio-Bistrom, M., Nelson, S. and Kristjanson, P. (2015). *Understanding gender dimensions of agriculture and climate change in smallholder farming communities. Climate and development*. Available at: www.tandfonline.com/doi/full/10.1080/17565529.2015.1050978?scroll=top&needAccess=true [Accessed 12 November 2016].

Kabeer, N. (1999). Resources, agency, achievements: Reflections on the measurement of women's empowerment. *Development and Change*, 30, pp. 435–464.

Kandiyoti, D. (1988). Bargaining with patriarchy. *Gender and Society*, 2(3), pp. 274–290. (Special Issue to Honour Jessie Bernard).

Kim, H. and Sambrook, C. (2014). *The changing role of women in the economic transformation of family farming in Asia and the Pacific*. Rome, Italy: International Fund for Agricultural Development. Available at: www.ifad.org/documents/10180/61297bc5-2280-4381-96b9-330b33a382bb [Accessed 17 June 2016].

Meinzen-Dick, R., Johnson, N., Quisumbing, A., Njuki, J., Behrman, J., Rubin, D., Peterman, A. and Waitanji, E. (2011). *Gender, assets, and agricultural development programs: A conceptual framework*. CAPRi Working Paper No. 99. Washington, DC: International Food Policy Research Institute.

Ministry of Agriculture. (2015). *Agricultural statistics at a glance 2014. Government of India*. New Delhi: Oxford University Press.

Mucavele, S. (2013). *The role of rural women in agriculture*. Available at: www.wfo-oma.com/women-in-agriculture/articles/the-role-of-rural-women-in-agriculture.html [Accessed 15 July 2016].

Ngenzebuke, R. (2014). *The returns of "I do": Women decision-making in agriculture and productivity differential in Tanzania*. Available at: www.worldbank.org/content/dam/Worldbank/Feature%20Story/Africa/afr-rama-lionel-ngenzebuke.pdf [Accessed 18 July 2016].

Planning Commission. (1952). *Five year plan*. New Delhi: Planning commission. Available at: https://arrow.latrobe.edu.au/store/3/4/2/8/0/public/B14218331.pdf [Accessed 10 June 2016].

Primo, N. (2003). *Gender issues in the information society*. UNESCO Publications for the World Summit on the Information Society.

Quisumbing, A. and Pandolfelli, L. (2010). Promising approaches to address the needs of poor female farmers: Resources, constraints, and interventions. *World Development*, 38(4), pp. 581–592.

Sanghi, S., Srija, A. and Vijay, S. (2015). Decline in rural female labour force participation in India: A relook into the causes. *VIKALPA The Journal for Decision Makers*, 40(3), pp. 255–268.

Singh, N.P., Singh, R.P., Kumar, R., Vashist, A.K., Khan, F. and Varghese, N. (2011). Adoption of resource conservation technologies in indo-gangetic plains of India: Scouting for profitability and efficiency. *Agricultural Economics Research Review*, 24, pp. 15–24.

Sultana, F. (2011). Suffering for water, suffering from water: Emotional geographies of resource access, control and conflict. *Geoforum*, 42, pp. 163–172.

UNDP. (1993). *Human development report 1993*. New York: Oxford University Press. Available at:http://hdr.undp.org/sites/default/files/reports/222/hdr_1993_en_complete_nostats.pdf[Accessed 10 June 2016].

Index

Note: Numbers in **bold** indicate a table. Number in *italics* indicate a figure.